THE
GLOBAL
VILLAGE

COMMUNICATION AND SOCIETY
edited by George Gerbner and Marsha Siefert

IMAGE ETHICS
The Moral Rights of Subjects
in Photographs, Film, and Television
Edited by Larry Gross, John Stuart Katz,
and Jay Ruby

CENSORSHIP
The Knot That Binds Power and Knowledge
By Sue Curry Jansen

THE GLOBAL VILLAGE
Transformations in World Life and Media in
the 21st Century
By Marshall McLuhan and Bruce R. Powers

SPLIT SIGNALS
Television and Politics in the Soviet Union
By Ellen Mickiewicz

TARGET: PRIME TIME
Advocacy Groups and the
Struggle over Entertainment Television
By Kathryn C. Montgomery

TELEVISION AND AMERICA'S CHILDREN
A Crisis of Neglect
By Edward L. Palmer

PLAYING DOCTOR
Television, Storytelling, and Medical Power
By Joseph Turow

THE GLOBAL VILLAGE

Transformations in World Life and Media in the 21st Century

MARSHALL McLUHAN
AND
BRUCE R. POWERS

New York   Oxford
OXFORD UNIVERSITY PRESS
1989

Oxford University Press

Oxford   New York   Toronto
Delhi   Bombay   Calcutta   Madras   Karachi
Petaling Jaya   Singapore   Hong Kong   Tokyo
Nairobi   Dar es Salaam   Cape Town
Melbourne   Auckland

and associated companies in
Berlin   Ibadan

Published by Oxford University Press, Inc.,
200 Madison Avenue, New York, New York 10016

Oxford is a registered trademark of Oxford University Press

Library of Congress Cataloging-in-Publication Data
McLuhan, Marshall, 1911–
The global village : Transformations in world life and media in
the 21st century / Marshall McLuhan and Bruce R. Powers.
p.   cm.
Bibliography: p.
Includes index.
ISBN 0–19–505444–X
1. Mass media—Technological innovations.
2. Mass media—Social aspects.
3. Technology—Social aspects.
I. Powers, Bruce R.
II. Title.
P96.T42M38   1989
302.2'34—dc 19   88–22718
CIP

9 8 7 6 5 4 3 2 1

Printed in the United States of America
on acid-free paper

Princeps gloriosissime, Michael, Archangele,
esto memor nostris; hic et ubique semper precare
pro nobis Filium Dei.

(Glorious Prince Archangel Michael, remember
us, here and everywhere, always intercede
for us (with) the Son of God . . .)

Is it a fact . . . that, by means of electricity, the world of matter has become a great nerve, vibrating thousands of miles in a breathless point of time? Rather, the round globe is a vast head, a brain, instinct with intelligence! Or, shall we say, it is itself a thought, nothing but a thought, and no longer the substance which we deemed it!

NATHANIEL HAWTHORNE
(*The House of the Seven Gables*)

Preface

I

Marshall McLuhan and I constructed this book from two points of view: the aesthetic and the technologic. Chapters 1 to 6 are an aesthetic meditation on how Marshall arrived at the tetrad through art and rhetoric. Chapters 7 to 9 concentrate on electronic communication technologies and demonstrate how speed-of-light technologies could be used to postulate possible futures. One might determine the final end of each electronic technology by its intensive enlargement or amplification. The four phases of the tetrad manifest the cultural life of an artifact in advance (whether it be a computer, data-base device, satellite, or global media network) by showing how a total saturated use would produce a reversal of the original intent.

To McLuhan, graphing the human use of an artifact could predict what society might do with a new invention. Hence, one could accept or reject from the beginning the future effects of any new artifact. For example, had a tetrad been constructed of the total human effects of atomic energy, we might have deployed all our secret services during World War II to frustrate the use of the atom as a weapon for any combatant, including ourselves. In more recent times, we might have foreseen that the effects of the contraceptive pill would create a deep birthrate plunge in many Western societies.

McLuhan believed an investigation of this book's precepts, his last collaborative work, would prove out his most profound thought: that the extensions of human consciousness were projecting themselves into the total world environment via electronics, forcing humankind into a robotic future. In other words, man's nature was being very rapidly translated into information systems which would produce enormous global sensitivity and no secrets. As usual, man was unaware of the transformation.

Because the present is always a period of painful change, every generation views the world in the past—Medusa is viewed through the polished shield: the rearview mirror. The Romans were obsessed with the world of Greece, the Greeks with the tribalists who preceded them (including that great primitive Socrates, whom Plato worshipped all his life). Plato did not have a clue as to what had brought literacy into the world or what it had done to philosophy. He spent his life as an amanuensis of Socrates, turning orality into an art form so as to cope with the new written literacy. But this is normal. People spend their lives making reasonable simulations of what has been done in the preceding age. The Renaissance man lived in the Middle Ages, mentally and imaginatively—deeply thrust through with uncritical classicism. The nineteenth-century man lived in the Renaissance. *We* live in the nineteenth century. The image that we have of ourselves, collectively, in the Western world is from that period. Tom Wolfe looks like a reconstituted Horace Greeley. Sherlock Holmes reigns on public television as an encyclopedic hero, a posture he would not have achieved in Victorian England. The typical American suburbanite lives in the frontier world of the nineteenth century; for him, Luke Skywalker is simply Billy The Kid revisited.

What is happening at the present time is that changes are occurring so rapidly that the rearview mirror does not work anymore— at jet speeds, rearview mirrors are not very useful. One must have a way of anticipating the future. Humankind can no longer, through fear of the unknown, expend so much energy translating anything new into something old but must do what the artist does: develop the habit of approaching the present as a task, as an environment to be discussed, analyzed, coped with, so that the future may be seen more clearly.

The tetrad replays various futures; it suggests experimental alternatives. The tetrad can therefore shift our perceptive focus from the past into the present. Take, for example, the book. Xerox makes it possible for every person to become his or her own publisher. We no longer need to mechanically print, repetitively, a particular text with little change. We can make a book to which people can constantly add pages—if necessary, from other books. Add the electronic data base for exploration and one might have access to the most unlikely combinations. Unlikely combinations produce discovery. *The Global Village* is not a nineteenth-century book, one of encyclopedic expectations; it is a book which never has the final answer, which brings the past into the present for the purpose of seeing an alternative future, a future where the whole of the economy appears to be moving rapidly toward tailor-made, individually committed services.

Marshall McLuhan, in his final years, wanted to talk to a new generation, one which was twenty to twenty-five years beyond *Understanding Media: The Extensions of Man* (1964). He said that the sons and daughters of the "Flower Children" would transform the world because they would find words to translate what had been ineffable to their parents.

The ineffable to McLuhan was what was dimly seen by those at Woodstock and Haight-Ashbury—that the entire world was in the grasp of a vast material and psychic shift between the values of linear thinking, of visual, proportional space, and that of the values of the multi-sensory life, the experience of acoustic space. Culturally, what is happening now is titanic. It needs a completely new frame of reference. McLuhan provides one. He presents it in a triad of new terms: visual space, acoustic space, and the tetrad. *The Global Village* is devoted to defining and explicating all three as it shows how world culture is repositioning itself to accept a completely different perceptive mode—the mode of the dynamically many-centered.

Visual space is the mind-set of Western civilization as it has proceeded over the last 4000 years to sculpt its monolithic linear self-image—a self-image which emphasizes the operation of the left hemisphere of the brain and which, in the process, glorifies quantitative reasoning.

Acoustic space is a projection of the right hemisphere of the human brain, a mental posture which abhors priority-making and labels and emphasizes the pattern-like qualities of qualitative thinking. McLuhan pointed out repeatedly that the passion of the visual space mind-set leaves little room for alternatives or participation. When no provision, for example, is made for two entirely different points of view, the result is violence. One person or another loses his identity. Acoustic space is built on holism, the idea that there is no cardinal center, just many centers floating in a cosmic system which honors only diversity. The acoustic mode rejects hierarchy; but, should hierarchy exist, knows intuitively that hierarchy is exceedingly transitory.

McLuhan espoused oriental values as primarily acoustic. Encyclopedic visual space is a mode developed by Plato, polished by Aristotle, and injected wholesale into Western thinking. The two value systems have interpenetrated each other for centuries, certainly when passed from hand to hand in slow print form. But, now the acoustic and the visual are separately slamming into each other at an explosive speed of light. Electric flow has brought differing societies into abrasive contact on a global level, occasioning frequent worldwide value collisions and cultural irritation of an arcing nature, so that, for instance, when a hostage is taken in Beirut an entire nation on the other side of the world is put at risk. McLuhan said, "in the last half of the 20th Century the East will rush Westward and the West will embrace orientalism, all in a desperate attempt to cope with each other, to avoid violence. But the key to peace is to understand both systems simultaneously."

Simultaneous understanding, or "integral awareness," can be seen in the tetrad. McLuhan invented the tetrad as a means of assessing the current cultural shift between visual and acoustic space. At present, every artifact of man mirrors the shift between these two modes.

In this book, we present a model for studying the structural impact of technologies on society. This model emerged from a discovery that all media and technologies have a fundamentally linguistic structure. Not only are they like language but in their essential form they *are* language, having their origins in the ability of man to extend himself through his senses into the environment.

Our research, at the Centre for Culture and Technology in Toronto, constituted an inquiry into the formal aspects of (linguistic) communication which, in the process, uncovered a tetradic structure: all media forms (a) *intensify* something in a culture, while, at the same time, (b) *obsolescing* something else. They also (c) *retrieve* a phase or factor long ago pushed aside and (d) undergo a modification (or *reversal*) when extended beyond the limits of their potential. The result is a four-part metaphor.

When this four-part "structure of the word" (*logos*) is applied to technologies one is able to ascertain the dynamic and social impact of any human artifact on the society into which it is extended; this can be formulated in a simple four-part analysis which is inclusive and, apparently, irreducible. In *The Global Village* we have confined our most widespread analysis to the approaching worldwide impact of video-related technologies which, in their present guise, forecast the most foreseeable future.

For Marshall McLuhan, the meaning of meaning was relationship. In the writing of this book from 1976 to 1984, I was swept into the ideational whirl of his family and associates. In the late forties, Marshall debated poetics with Ezra Pound at Saint Elizabeth's and, through letters, drifted into an intense exchange of critiques concerning Pound with others, such as Hugh Kenner and Felix Giovanelli. In a similar way, I was caught up in a quick-fire exchange of analytical facts and opinions with Marshall, his friends, and colleagues, both in Toronto and elsewhere. McLuhan and I talked; made critical tapes on our ideas; and revised commonly circulated preparatory texts, especially on the structuring of the tetrad. Marshall would take the same ideas and share them with such luminaries as Glenn Gould, John Cage, and Pierre Trudeau. Constant refinement through the minds of others was his way of working. In discussing his penchant for sharing his ideas in development with whoever would listen, McLuhan once told Eric McLuhan and myself—Eric, who was so important in later years in helping his father to record ideas and conversations for revision—"Truth is not matching. It is neither a label nor a mental reflection. It is something we make in the encounter with the world that is making us. We make sense not in cognition, but in replay. That is my definition of intellection, if not, indeed, scholarship. Representation, not replica."

II

In the weeks just before his final stroke in 1979, McLuhan was preoccupied with death. The thought had sprung out of our discussions about the central metaphor of *Understanding Media,* the Narcissus myth.

One Saturday morning, in examining the opening of our book, *The Global Village,* Marshall noticed the relationship between the astronauts' first view of the earthrise (see Chapter 1) and the mirror image percept he had first examined in 1963. When we went to the moon, he said, we expected photographs of craters; instead, we got a picture of ourselves. Ego trip. Self-love.

The mirror image, I countered, is another way of saying water, which stands for change in man and nature. Narcissus fell in love with his image in the water. "No," said Marshall, "that's the popular conception." Narcissus, as Ovid paints him, is a primal youth, has never seen a mirror or his image. "He fell in love with someone else." That's the mythical and satirical point. For him, the watery mirror was death.

Marshall paused and walked across the living room to place a log on the fire. "Had you thought of the nature of hell in ancient Near Eastern literature?" I asked Marshall. Hell is a watery place. Remember Gilgamesh. The Bible refers to it as Sheol. The Greek shades wander in a dark and misty underworld. By concentrating only on the watery image, Narcissus performs a sort of dreamlike closure, Marshall said. Eventually, like Alice, he must pass through the vanishing point, to see both sides of the mirror. Marshall seemed thunderstruck. "That's what death must be like; one sees oneself simultaneously, as oneself and as the other." Like seeing your own face, warts and all, on the TV screen for the first time. The actor without his makeup. The anchorman without his hairpiece. Christ walks on the water. Peter falls through it. Water is death for humans and a container for the diabolic.

"The diabolic?" I said. From a Christian viewpoint, the Devil brought death into the world. When we outer ourselves, we are in a saving community, the realm of consciousness. When we are innering, within ourselves, through the mirror as it were, we are in danger of being lost in the funhouse of our unconscious. Trapped inside one's skull. A definition of insanity? It's quite explicit in the Gospels: "Unless you die again, you cannot be reborn." That's Poe's maelstrom; into the vortex and out again, surviving not only because you travel light but are prepared to jettison everything. One dies and is reborn. One is immersed and rises again. The cross of the pagan: he cannot return from hell. Judgment Day for the Christian is seeing himself on earth and in the hereafter simultaneously, which is the unique characteristic of speed-of-light technologies. "When you see yourself on TV, as I have, you are innering and outering simultaneously." A diabolical plot? (Later, McLuhan's letters would reveal that he told Jacques Maritain that the Prince of this world must be a great electrical engineer.) We concluded that video-related technologies might produce a form of psychological death for all mankind by separating it permanently from the natural order, the book of nature, through Narcissus-like self-involvement, a conclusion reached by McLuhan operating on three analytical levels at once: the perceptual, the historical and the analogic. That was McLuhan's rhetorical style, to explore, and re-explore, a subject with a myriad of ideas, each seeming to have an equal judgmental weight, rather than a single point of view. I hope this explanation is of some help to the first-time reader. *The Global Village,* after all, is the first right-hemisphere book as well as McLuhan's last work.

As I review the copyedited manuscript I am pleased that so many of our tetradic projections are as relevant in 1988 as they were when *The Global Village* was being put together (from 1976 to 1984). Yet new technologies are arising which beg for analysis, like cellular phone systems and digitally controlled 360-degree film projection processes. But that's another book.

Lewiston, New York   B. R. P.
August 8, 1988

Contents

EXPLORATIONS
IN VISUAL AND
ACOUSTIC SPACE

The Resonating Interval

All Western scientific models of communication are—like the Shannon-Weaver model—linear, sequential, and logical as a reflection of the late medieval emphasis on the Greek notion of efficient causality. Modern scientific theories abstract the figure from the ground. For use in the electric age, a right-brain model of communication is necessary to demonstrate the "all-at-onceness" character of information moving at the speed of light. As voice, print, image, and sensory data proceed simultaneously, figure and ground are often in apposition rather than in a sequential relationship. For example, the consciousness of the data-base user is in two places at once: at the terminal and in the center of the system.

An artifact pushed far enough tends to reincorporate the user. The Huns lived on their horses day and night. Technology stresses and emphasizes some one function of man's senses; at the same time, other senses are dimmed down or temporarily obsolesced. The process retrieves man's propensity to worship extensions of himself as a form of divinity. Carried far enough man thus becomes a "creature of his own machine."

The trick is to recognize the four-fold pattern of transformation before it is completed. At full maturity the tetrad reveals the metaphoric structure of the artifact as having two figures and two grounds in dynamic and analogical relationship to each other. The resonating interval defines the relation between figure and ground and structures the configuration of ground. Through comprehensive awareness we may see both past and future at once. Strictly left-brain thinking, or "angelism," allows technology to move as a dumb force because without perceiving all four-fold processes in operation, we are unconscious of their overall effects.

After the Apollo astronauts had revolved around the moon's surface in December of 1968, they assembled a television camera and focused it on the earth. All of us who were watching had an enormous reflexive response. We "outered" and "innered" at the same time. We were on earth and the moon simultaneously. And it was our individual recognition of that event which gave it meaning.

A resonating interval had been set up. The true action in the event was not on earth or on the moon, but rather in the airless void between, in the play of the axle and the wheel as it were. We had become newly aware of the separate physical foundations of these two different worlds and were willing, after some initial shock, to accept both as an environment for man.

The same may be said for the left and right brain.[1] We must once again accept and harmonize the perceptual biases of both and understand that for thousands of years the left hemisphere has suppressed the qualitative judgment of the right, and the human personality has suffered for it. The isolation and amplification of one sense, the visual, is no longer enough to deal with acoustic conditions above and below the surface of the planet.

The book of nature contains innumerable borderlines and interfaces. The resonant interval may be considered an invisible borderline between visual and acoustic space.[2] We all know that a frontier, or borderline, is a space between two worlds, making a kind of double plot or parallelism, which evokes a sense of the crowd, or universality. Whenever two cultures, or two events, or two ideas are set in proximity to one another, an interplay takes place, a sort of magical change. The more unlike the interface, the greater the tension of the interchange.

The tetrad, like the metaphor, performs the same function that the camera did in the *Apollo 8* mission: it reveals figure (moon) and ground (earth) simultaneously.[3] The left brain with its sequential, linear bias, hides the ground of most situations, making it subliminal. Left-hemisphere thinking, as a dominant mode, is linear and tends to place emphasis only on the connected; it is steeped in a priori notions of order, masking the complementarity of both right and left brain modes.

The terms *figure* and *ground* were borrowed from gestalt psychology by the Danish art critic Edgar Rubin, who about the year 1915 began to use them to discuss the parameters of visual perception.[4] At the Centre for Culture and Technology, we broadened Rubin's usage to take in the whole of perception and consciousness. All cultural situations are composed of an area of attention (figure) and a very much larger area of inattention (ground). The two are in a continual state of abrasive interplay, with an outline or boundary or interval between them that serves to define both simultaneously. As in the paintings of Van Gogh or cloisonné art, figures rise out of and recede back into ground, which is configurational and comprises all other (available) figures at once. For example, at a lecture attention will shift from the speaker's words to his gestures, to the hum of the lighting or street sounds, or to the feel of the chair or a memory or association or smell, each new figure alternately displacing the others into ground.

One sees a ready comparison with Gombrich's description of *synesthesia*.[5] Writing in *Art and Illusion,* E. H. Gombrich sees the interplay of sensory input and response as a kind of mosaic, a configuration:

> What is called "synesthesia," the splashing over of impressions from one sense modality to another, is a fact to which all languages testify. They work both ways—from sight to sound and sound to sight. We speak of loud color or of bright sounds; and everyone knows what we mean. Nor are the ear and the eye the only senses that are thus converging to a common center. There is touch in such terms as velvety voice and a cold light, taste with sweet harmonies of color and sounds. . . .

The common sensorium, which is Goethe's proper use of the word *Weltinneraum,* contains all potential figures in sensuous latency at once. In this respect, ground provides the structure of or style of awareness, the way of seeing or the terms on which a figure is perceived. The study of ground on its own terms is virtually impossible, as by definition it is at any moment environmental and subliminal. The only possible strategy is to construct an anti-environment, which is the normal activity of the artist, the only person in the culture whose whole business is the retraining and updating of sensibility.

In the order of things, ground comes first. The figures arrive later. Coming events cast their shadows before them. The ground of any technology is both the situation that gives rise to it as well as the whole environment (medium) of services and disservices that the technology brings with it. These are the side effects, and they impose themselves haphazardly as a new form of culture. The medium is the message. As an old ground is displaced by the content of the new situation, it becomes available to ordinary attention as figure. At the same time a new nostalgia is born. The business of the artist has been to report on the nature of ground by exploring the forms of sensibility made available by each new ground, or mode of culture, long before the average man suspects that anything has changed.

Audile (acoustic) space and tactile (visual) space are in fact inseparable. But in the interfaces created by these senses, figure and ground are in dynamic equilibrium, each exerting pressure on the other across the interval separating them. The interface therefore, is resonant and not static. That pressure creates a condition of continual, potential transformation called *chiasmus*. Resonance is the mode of acoustic space; tactility is the space of the significant bounding line and of interval.

The tetrad, taken as a whole, is a manifestation of human thinking processes. As an exploratory probe, tetrads do not rest on a theory but a set of questions; they rely on empirical observation and are thus testable. When applied to new technologies or artifacts, they afford the user predictive power; in this sense as well they may be viewed as a scientific instrument. Once again, insofar as the tetrads are a means of focusing awareness of hidden or unobserved qualities in our culture and its technologies, they act phenomenologically. From Hegel to Heidegger, phenomenologists have engaged in an attempt to get at the hidden properties, or concealed effects, of language and technology alike. To do so they have tackled a right-hemisphere problem using left-hemisphere techniques and modes of cognition—which is comparable to tap-dancing in chains! The tetrad offers a way out of this dilemma.

Until now, the conventional form in analysis or exposition has been triadic and logical, as in the syllogism. It is a propositional left-hemisphere form, rigid and connected, in the pattern of efficient cause.[6] Whether syllogistic or Hegelian-dialectical, for some inherent reason the triad eliminates ground. When a fourth term is added, the structure becomes resonant and appositional and metaphoric: simile, metonymy, synecdoche yield to metaphor.

The tetradic representation of processes had led us to an awareness that all our artifacts are in fact words. All of these things are the outerings and utterings of man. In Douglas Fraser's *African Art as Philosophy,* it is mentioned as a feature of some traditional societies that speech and weaving are synonymous:

> Among the Bambara and the Dogon, the gift of weaving is closely associated with that of speech. *Soy* the Dogon word for cloth means "it is the spoken word" (Griaule, 1948, p. 30). Weaving, along with speech, was a gift from the creator to help man. . . .

Each tetrad is the word or the *logos* of its subject, and all these words are peculiarly human, with the utterer as the etymology. They constitute, in opposition to the Shannon-Weaver construct, a right-hemisphere theory, or model, of communication; and, as they provide both exegesis and etymology of a (rhetorical) utterance, they serve to bring up to date the ancient and medieval tradition of grammar-allied-to-rhetoric in a way that is consonant with the forms of awareness imposed on the twentieth century by electronic technology.

They are equally applicable to the full range of human artifacts whether hardware (objects) or software (ideas), although our left-hemisphere training makes them easier to apply to the former than to the latter. They provide an analytic of their subjects from the standpoint of *logos* and formal cause. Just as all artifacts are words, all words and languages are artifacts; each of which manifests a four-part structure in the form of double-ends joined. There appear to be no exceptions. This is the right-hemisphere aspect of language. All nonverbal objects, whether safety pins or ICBMs, including also the laws of science and institutions, share this same four-part *logos*-structure in their manifestations and effects.

The tetradic metaphor opens up to the matter of the grammar and syntax of each artifact. There seem to be only four features, and they are in analogical proportion to each other. The role of the metaphor is the elevation of hidden ground into sensibility. For example, "hearts of oak," where the hidden ground is "hearts of our people." A double figure-ground relation is established so that: "ordinary hearts are to these hearts as ordinary wood is to oak," and the complementary structure also applies: "ordinary hearts are to ordinary wood as these hearts are to oak."

As we have said before, technologies, like words, are metaphors.[7] They similarly involve the transformation of the user insofar as they establish new relationships between him and his environments. A double figure-ground relationship is brought into play with "natural man is to man-with-artifacts as is the natural environment to the man-made environment." And the complementary, "natural man is to the natural environment as is man-with-technology to the man-made environment."

The parts of the tetrad have the same complementary character:

Retrieval is to obsolescence as enhancement is to reversal
—and—
Retrieval is to enhancement as obsolescence is to reversal.

The relationship of elements reflected in the metaphor is another way of saying that the left and right brain may be capable of interchange, but are nevertheless incommensurable. The left hemisphere places information structurally in visual space, where things are connected sequentially—having separate centers with fixed boundaries. On the other hand, acoustic space structure, the function of the right brain in which processes are related simultaneously, has centers everywhere with boundaries nowhere. The former is like a painting or a photograph in perspective. The latter may be likened to a symphonic surround.

The left and right brain relate yet lack a common basis for comparison. Simultaneous interplay cannot be reduced to linear (sequential) representation in much the same way that a synchronic chord of music cannot be experienced as a diachronic tune. There is evidence that the whole brain functions more like a hologram[8] than a one-thing-at-a-time computer; and, in the same sense, every human artifact is a medium of communication whose message may be said to be the totality of the satisfactions and dissatisfactions it engenders which, at the speed of light, reveal simultaneous process patterns. To arrive at this set of process patterns, one asks the questions:

1. What does any artifact enlarge or enhance?
2. What does it erode or obsolesce?
3. What does it retrieve that had been earlier obsolesced?
4. What does it reverse or flip into when pushed to the limits of its potential (chiasmus)?

As framed, the tetradic metaphor amplifies the potential equilibrium of the relations being explored; it obsolesces simile, metonymy, and connected logic; it retrieves understanding, or meaning, by virtue of replay in another mode; and it reverses into allegory or parallelism (see Fig. 1.1).

The tetrad, as a right-hemisphere visualization, helps us to see both figure and ground at a time when the latent effects of the mechanical age tend to obscure the ground subliminally.[9] Its chief utility is that it raises the hidden ground to visibility, enabling the analyst to perceive the double action of the visual (left hemisphere) and the acoustic (right hemisphere) in the life of the artifact or idea. As such, the tetrad performs the function of myth in that it compresses past, present, and future into one through the power of simultaneity. The tetrad illumines the borderline between acoustic and visual space as an arena of spiraling repetition and replay, both of input and feedback, interlace and interface in the area of an imploded circle of rebirth and metamorphosis.

The action of any artifact (or its corresponding idea) is diachronic as it undergoes a progressive history and development from enhancement—which should be regarded as a form of amplification—to obsolescence (A to B to D to C). It is synchronic if one were to view the artifact mythically as a configuration (A/D = C/B and B/D = C/A). When the artifact manifests itself at high relief, which means its development has been nominally revealed, all its process patterns may be said to be simultaneous, like an electrical circuit. The balanced tetrad possesses two grounds and two figures in ratio to each other.[10]

The Greeks and the Romans invented the historical sense (the diachronic) so that it would be possible to carve up and deal with time as a rational control device. Emphasizing figure to the virtual

Tetrad Structure

A. Enhancement
(figure)

D. Reversal
(ground)

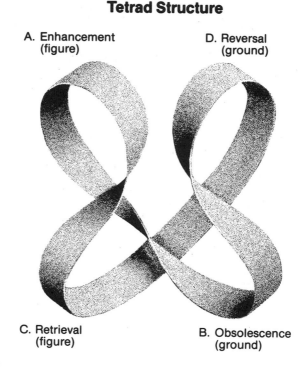

C. Retrieval
(figure)

B. Obsolescence
(ground)

Fig. 1.1.

exclusion of ground is a recent Western innovation—mostly nineteenth-century. Yet large areas of the planet still perceive time as circular or synchronic, testifying to the idea that human experience may occur in nothing but the present. Such a duality of attitudes suggested to T. S. Eliot that past, present, and future exist as one thing: time considered as sequential (left hemisphere) is figure and time considered as simultaneous (right hemisphere) is ground.

The tetrad not only reveals the configurational character of time, but also that the artifact (or founding idea) is always the product of the user's mentality. The tetrad includes the ground of the user, as utterer; and paradoxically, includes the user as ground. We make ourselves and what we make is perceived as reality. For example, an analysis of the effects of the printed word on another environment usually engenders quite different results. The tetrads for print in the United States, China, or Africa would have three different grounds.

The tetrad helps us to see "and-both," the positive and the negative results of the artifact. For instance, the automobile amplified one's ability to cover distance more quickly and, to a limited extent, carry cargo. Yet, almost from its beginning, this invention simultaneously affected man's relationship to time and space, obsolescing the forms of social organization rooted in pedestrian and equestrian traditions. The township and the neighborhood collapsed. The inner city was left to nonhuman-scale development, while that space in the city that had been set aside as human-size living space was shifted to the suburbs.

The gasoline automobile brought back a sense of private identity and independence which had first manifested itself on the American frontier and, to a lesser extent—as Mark Twain tells us—in the social threads of the farm and village. Pushed to an extreme, in urban sprawl, congestion, and pollution, the automobile reverses into the electric mini-car and encourages renewed activity in jogging, bicycling, and urban nature preserves.

Even before the Organization of Petroleum Exporting Countries' price squeeze, over-amplification had made the automobile a monster. When the figure (car) is on the verge of swallowing the ground (environment) it becomes grotesque. Tribalism as the ground of pre-literate humanity, as Nevitt puts it, was a prodigy of "rhyme without reason" which, before the Greek city-states, allowed no private identity. Civilization, the artifact of literate man, when pushed to its farthest limits trumpets "reason without rhyme" and makes a meal of all humanity. The tetrad could serve us by revealing the all-embracing system—whether of a monolithic state or a well-intentioned business monopoly—before the figure-ground interface is erased.

The new video-related technologies promise to impose a new monopoly of ground over figure. Whatever is left of mechanical age values could be swallowed up by information overload. Media determinism, the imposition willy-nilly of new cultural grounds by the action of new technologies, is only possible when the users are well-adjusted, i.e., sound asleep. The vortex of side effects was penned by James Joyce: "willed without witting, whorled without aimed." There is no inevitability, however, where there is a willingness to pay attention.

The tetradic analysis is a way of anticipating the changes in the *ma* (negative space); or, of anticipating and perceiving the *ma* as part of the overall configuration—as a complete entity—rather than of restricted and fragmented portions. The impetus of change is like the force of excited motion in the atom. It may proceed at great velocity but eventually returns to its lowest state. Nothing essential has been lost; simply a metamorphosis of mass into energy and vice versa. Technological enlargement is a process toward excess. As part of his spiritual health, man should make as his first object the recognition of pattern as a means to avoid excess and achieve equilibrium. As Aristotle suggests in *De Anima;* it is inherent in the psychologic structure of the artifact. It is accomplished only by conscious choice. Angelism, sometimes called discarnatism, allows technology to move as a dumb force, because without perceiving all four-fold processes we are unconscious of its effects. Discarnatism floats in the abstract clouds, without any relation to ground, or environment—the besetting sin of academic hypothesis.

The Wheel and the Axle

Touch is the resonant interval or frontier of change and process, and is indispensable for the study of technological effects. Interface is the basis of the relationship between visual and acoustic space. Much of present-day confusion and ferment stem from the divergent experience of Western literate man on the one hand, and his new surround of simultaneous and acoustic knowledge on the other. Part of this problem stems from an inadequate understanding of the nature of the archetype. An archetype has both an overt and hidden side (figure and ground). The tetrad reveals both. The hidden effects of any archetypal situation are the aspects which truly shape our behavior.

Many academic fields of inquiry have been stymied by a misunderstanding of how visual and acoustic space relate to the notions of the diachronic and synchronic. This confusion mirrors the whole disjointedness of Western education, which emphasizes left-brain over right-brain cognition; and, which may be traced back to the writings of both Plato and Aristotle. The tetradic analysis corrects this imbalance.

The idea of interface, of the resonant interval as "where the action is" in all structures, whether chemical, psychic, or social, involves the factor of touch. Touch, as the resonant interval or frontier of change and process, is indispensable to the study of structures. It involves also the idea of play as in the action of the interval between the wheel and the axle. Play literally constitutes the basis of human communication since human beings do not match ideas so much as reinterpret them.

Electronic man, having found himself in an arena of simultaneous information also finds himself increasingly excluded from the older more traditional (visual) world in which space and reason seem to be uniform, connected, and stable. Instead, Western (visual and sequential) man now discovers himself habitually relating to information structures which are simultaneous, discontinuous, and dynamic. He has been plunged into a new form of knowing, far from his customary experience tied to the printed page. In the same way that the sense of hearing apprehends details from all directions at once, within a 360-degree sphere, as it were, in a manner similar to a magnetic or electrical field; so knowing itself is being recast and retrieved in acoustic form. As such, by the next century it will destroy all existing forms of school structures. "Back to basics" is the last bugle call of the diehards.

In 1919 T. S. Eliot in his essay "Tradition and the Individual Talent" had stressed the view that all art from Homer to the present formed a simultaneous order and that this order was perpetually motivated, renewed, and retrieved by new experience. His symbolist approach to language and art and communication is well indicated in his celebrated definition of the historical sense in poetry:

> The existing monuments form an ideal order among themselves, which is modified by the introduction of the new (the really new) work of art among them. The existing order is complete before the new work arrives; for order to persist after the supervention of novelty, the *whole* existing order must be, if ever so slightly, altered; and so the relations, proportions, values of each work of art toward the whole are readjusted; and this is conformity between the old and the new.[1]

This definition points to the endless process of change and transformation and retrieval implicit in the simultaneous and homeostatic character of art, which has as its ultimate end a dedication to eternal stability. Much of the confusion of our present age stems intrinsically from the divergent experience of Western literate man, on the one hand, and his new surround of simultaneous or acoustic knowledge, on the other hand. Western man is torn between the claims of visual and auditory cultures or structures. Visual culture is fragmented; acoustic culture is integrated (see Chapter 3).

Neo-acoustic space, centering on electronic technologies, gives us simultaneously access to all pasts. As for tribal man we have said before there is no history: all is present, as the mundane unobtrusively changes into the mythic:

> If we can consider form the reversing of archetype into cliché, as, for example, the use of an archetypal Ulysses in James Joyce's novel to explore contemporary consciousness in the city of Dublin, then we may ask what would be the status of this pattern in primordial times, in the medieval period, and today. The answer would seem to be that in primordial times and today this archetype-into-cliché process is perfectly normal and accepted but that in the medieval period it is exceptional and unusual.

> The Balinese say, "We have no art, we do everything as well as possible." The artist in the Middle Ages, Renaissance, or the era up to the nineteenth century was regarded as a unique, exceptional person because he used an exceptional, unusual process. In primordial times, as today, the artist uses a familiar, ordinary technique and so he is looked upon as an ordinary, familiar person.

> Every man today is in this sense an artist—the administrator, the scientist, the doctor, as well as the man who uses paint or sculpts stone. Just as the archaic man had to follow natural processes of rhythms in order to influence and to purge, cleanse them by *ricorso,* so modern electric technologies require such timing and precision that only the following of processes in nature can be tolerated. The immediately preceding centuries of mechanization had been able to bypass these processes by fragmentation and strip-mining kinds of procedures.[2]

The fall or scrapping of our old visually oriented culture, so beloved of nineteenth-century oligarchs, puts us all into the same archetypal cesspool, engendering a nostalgia for earlier, more secure (because fixed in memory) conditions. The old Pierce Arrow seems inevitably better built and safer than the newest car. Keep in mind that, initially, any cliché is a breakthrough into a new dimension of experience.

Alfred North Whitehead mentions in *Science and the Modern World* that the great discovery of the nineteenth century was the detection of the technique of discovery. The art of discovery, of retrieval—that is, the use of acoustic, probing awareness as method—

is itself now a cliché, and the idea of creativity has become one of the major stereotypes of the twentieth century.

The archetype, which depends on an overarching comprehension of the past (the mythic milieu), is retrieved awareness or consciousness. It is consequently a retrieved combination of clichés—an old cliché brought back by a new cliché. Since a cliché is a unit of extension of man, an archetype is a quoted extension, medium or technology, or environment. Flapper skirts and cloche hats always quote the Roaring Twenties. The recent Broadway offering *A Chorus Line* quotes the frenetic musicals of the thirties but in a different context. The rising ingenue is submerged in a galaxy of would-be stars: cloned Ruby Keelers and Dick Powells. A new twist on an old tune.

The following are examples of archetypes which have been chosen to stress the normal tendency of a cliché to cross-quote from one technology to another, in the sense of an established technology being mutated by a newer one:

> a flagpole flying a flag (spear and banner)
> a cathedral adorned with a stained-glass window
> pipeline carrying oil
> cartoon with a caption
> story with an engraved illustration
> perfume advertisement with a sachet of perfume
> an electric circuit feeding an electric log fire
> ship with a figurehead
> a television set being used as a bulletin board

A flagpole flying a modern flag may become a complex retrieval system. The flag could be a Russian flag, with its hammer and sickle. As flagcloth, the flag could retrieve an entire textile industry as to technique and fibers. By virtue of the fact that the flag is a national flag, it could retrieve flags of other nations. And so on.

The isolated cliché, in other words, is incompatible with other isolated single clichés, but the archetype is extremely cohesive; the residues of other archetypes adhere to it. When we consciously set out to retrieve one archetype, we unconsciously retrieve others; and this retrieval occurs in infinite regress. In fact, whenever we quote one consciousness, we also quote the archetypes we exclude; and this quotation of excluded archetypes has been called by Freud, Jung, and others "the archetypal unconsciousness."[3]

The Freudians and the analytical psychologists have had a tremendous impact on the educated modern temper. And it is for this reason alone that we must, at this point, devote some time toward clearing the air in regard to the nature of the archetype. Jung and his disciples have been careful to insist that the archetype is to be distinguished from its expression. Strictly taken, a Jungian archetype is a power or capacity of the psyche. Nevertheless even in Jung's writing the term is used with interchangeable meaning.

In *Psyche and Symbol,* Jung declares that "the archetype is an element of our psychic structure and thus a vital and necessary component in our psychic economy. It represents or personifies certain instinctive data of the dark primitive psyche: the real, the invisible *roots of consciousness*." Jung is careful to remind literary critics to consider the archetype as a primordial symbol:

> The archetypes are by no means useless archaic survivals or relics. They are living entities, which cause the praeformation of numinous ideas or dominant representations. Insufficient understanding, however, accepts these praeformations in their archaic form, because they have a numinous appeal to the underdeveloped mind. Thus communism is an archaic, primitive and therefore highly insidious pattern which characterizes primitive social groups. It implies lawless chieftainship as a vitally necessary compensation, a fact which can only be overlooked by means of a rationalistic one-sidedness, the prerogative of the barbarous mind.
>
> It is important to bear in mind that my concept of the "archetypes" has been frequently misunderstood as denoting inherited patterns of thought or as a kind of philosophical speculation. In reality they belong to the realm of the activities of the instincts and in that sense they represent inherited forms of psychic behaviour. As such they are invested with certain dynamic qualities which, psychologically speaking, are designated as "autonomy" and "numinosity."[4]

Jung accounts for his theory of archetypes by means of the hypothesis of a collective race memory, although he is well aware that there is no scientific acceptance for such an idea. His justification, however, for using the concept of a collective memory is based on the recurrence over a wide area of archetypal patterns in artifacts, literatures, the arts, et cetera, apart from its shaky scientific basis. In other words, Jung appeared to subscribe to the existence of key artifacts as a projection of man's consciousness, perhaps as his starting point.

Using McLuhan's tetrad, we may recapitulate by observing that as the new artifact form or technology pervades the host culture as a new cliché, it displaces, in the process, the old cliché or homeostasis to the cultural rag-and-bone shop, and older clichés are retrieved both as inherent principles that inform the new ground and new awareness, and as archetypal nostalgia figures in relationship to the new ground—all of which is accomplished with transformed meaning. The automobile ended the age of the horse and buggy, but the horse and buggy returned with new significance and experience in grade "B" Westerns. We saw that swaying gig in relation to our own experience with the horseless carriage.

The tetrads are an instrument for revealing and predicting the dynamics of innovations and new situations. The usual archetypal explanations are inadequate because they nevertheless regard the archetype as a figure minus its ground, like the New Guinea tribesman photographed by Irving Penn against an unnatural white background. The image is sharpened but the context is lost. In this regard, Jean Piaget observed:

> Before we go on, we should stress the importance of this notion of equilibration, which enables us to dispense with an archetypal explanation for the prevalence of good forms. Since equilibration laws are coercive, they suffice to account for the generality of such processes of form selection; heredity need not be called in at all. Moreover, it is equilibration which makes *Gestalten* reenter the domain of structure as circumscribed by us in Section 1, for whether physical or psychological, equilibration involves the idea of transformation within a system and the idea of self-regulation. Gestalt psychology is therefore a structuralist theory more on account of its use of equilibration principles than because of the laws of wholeness it proposes.[5]

Both the retrieval and reversal aspects of McLuhan's tetrad involve metamorphosis. The tired cliché of the movie becoming an art form when television became the dominant entertainment is transparently clear today. Likewise the entire planet has been subsequently retrieved through satellite, cum TV, as a programmable resource and art form (i.e., the notion of ecology) as a side effect of the new satellite surround, or ground. Once again, money obsolesces barter, or the swapping of unlike items, which may or may not be perishable. But it retrieves potlatch, or Indian blanket count, in the shape of conspicuous consumption. The digital watch replaces the old circular dial and retrieves the sundial form; which likewise, in ancient times, used light itself to tell the time and which also had no moving parts.

In the West, electronic technology displaces visual space and retrieves acoustic space in a new form, as the ground now includes the detritus of alphabetic civilization. The effect in the East is quite different, to the degree that Asian cultures put on Western clothes of phonetic alphabet and hardware. The alphabet becomes their means of transformation from group-think to individualism. Harold Innis examined the process whereby, through a shift in the media of writing, temple bureaucracies were displaced by military bureaucracies, and expansion, or conquest, programs began.[6] A few years ago, Iran was reeling under the impact of electronic media and was turning back from a military to a temple-control government under the Mullahs, spearheading a retrieval movement to archaic traditional mores that are more latent in many of Iran's neighbors, such as Iraq. The recent war with Iraq was a further acting out of that return to tribal values facilitated by the acoustic properties of electric media. Radio, loudspeakers, and audiocassettes brought the Mullah's cry to the force of a thunderclap, on a regional scale.

The reversal aspect of the tetrad is succinctly exemplified in a maxim from information theory: *data overload equals pattern recognition.*[7] Any word, or process or form, pushed to the limits of its potential, reverses its characteristics and becomes a complementary form, just as the airplane reverses its controls when it passes through the sound barrier. Any perceptive TV journalist realizes that a news item moved at electronic speeds acquires infinite mass. The resulting information overload is a world in which all patterns stand out loud and clear for the first time. Money as hardware, pushed to its limit, reverses into the lack of money— credit (software or information). At high speed or in great quantity, either on the test track or on the superhighway, the car reverts to nautical form, and traffic (as a crowd) "flows."

In the same sense, by repetition an archetype can become a cliché again, or an individual man can become a crowd. The cloned person loses his private identity, but becomes a corporate one. As we have said before, breakdown becomes breakthrough. At electronic speeds all forms are pushed to the limit of their potential: on the telephone (or on the air) it is not the message that travels at electronic speed. What actually occurs is that the sender is sent, minus a body, and all the old relationships of speaker and audience tend to be erased.

The dominant processes which rise to the surface in the tetrad form are intended to reveal some of the subliminal and previously inaccessible aspects of technology. To the extent that these observations reveal the hidden effects of artifacts on our lives, they are endeavors of art, bridging the worlds of biology and technology. Between the artifact and the personal or social response there is an interval of play as between the wheel and the axle. This interval constitutes the figure-ground gestalt of interaction and transformation.

H. J. Eysenck, the Dutch-born British psychologist, observes:

In some form or other, the law of effect has been one of the most widely recognized generalizations in the whole of psychology. "The belief that rewards and punishments are powerful tools for the selection and fixation of desirable acts and the elimination of undesirable ones." (Postman, 1947) is almost universal, and although the law itself is usually associated with the name of Thorndike (1911) who first used this phrase, he had precursors, e.g., Bain (1868) and Spencer (1870) who brought together the contributions of Associationism, Hedonism, and the Evolutionary doctrine in a coherent form closely resembling Thorndike's own formulation. This formulation was as follows: "of several responses made to the same situation, those which were accompanied or closely followed by satisfaction to the animal will, other things being equal, be more firmly connected with the situation, so that when it returns, they will be more likely to recur; those which are accompanied or closely followed by discomfort to the animal will, other things being equal, have their connection with the situation weakened so that, when it recurs, they will be less likely to occur. The greater the satisfaction or discomfort the greater the strengthening or weakening of the bond."[8]

A figure is an area of special psychic attention. The law of effect is strangely concentrated on the figure and its encounter with other figures, rather than the figure in relation to the ground, or the total situation. Western man, in his categorical mode, battens on one or two elements in any situation and represses the rest. Connections are visual: there is actually no connection between figure and ground but only interface. The left-hemisphere bias in Western thought which directs the attention to the figure or the idea or the concept is typical, not only of psychology but of philosophy and of science.

Anthropology, on the other hand, began by using the ground or the total culture itself as a figure for attention, thus seeming to break with the 2000-year tradition of considering figure-minus-ground. In Thomas Kuhn's book *The Structure of Scientific Revolutions,* the paradigms of continuous or extended metaphors, which he sees as channeling scientific endeavors in various times and fields, are considered as figures without any social or cultural ground whatever.[9] The only interplay that they are allowed in his work is with other paradigms, past or present.

Perhaps one reason why T. S. Eliot's "Tradition and the Individual Talent" was so revolutionary was that Eliot considered the totality of language and culture as a unified ground to which the individual talent had to relate. One of the basic assumptions of normal science is the left-brain need for measurement and quantification of effects.

The left-hemisphere paradigm of quantitative measurement and precision, recently re-explicated by some neurophysiologists, depends on a hidden ground which has never been thoroughly discussed by scientists in any field. That hidden ground is the acceptance of visual space as the norm of science and rational endeavor. Visual space alone can enclose, or be enclosed, and yet it was a by-product of the phonetic alphabet. The original implementors and users of visual space had, and have, the hidden phonemic ground of their discoveries or of their preferences in the organization of thought and exploration. Today it is easy for us to perceive what programmed them as a hidden ground, because that ground has itself become a figure starkly portrayed against the new ground of the electronic information environment. Instant information, as an environment, has the effect of pushing all other subliminal effects up into consciousness. That is, it has this effect with regard to all forms except itself, for the effect of the electronic environment is to turn people inward and to substitute the inner trip for outer exploration, being for becoming. That the hidden grounds of other cultures should now be available for inspection has created the investigations of structural linguistics and of anthropological and ecological studies on a world scale. For the structural is constituted by the simultaneous and is antithetic to the visual, which it now makes perceptible as an exotic figure. When the environment of instant electronic information becomes the hidden ground of all perception, choice, and preference, the ground which underlays the world of precise and quantifiable scientific study is pushed aside or dissolved.

All of our senses create spaces peculiar of themselves, and all of these spaces are indivisible and immeasurable. Tactile space is the space of the resonant interval, as acoustic space is the sphere of simultaneous relations. They are as indivisible as osmic or kinetic space (smell or stress).

The study of the law of effect has been the area of scientific study since Galileo; but when data became available at electronic speeds of retrieval, pattern recognition and transformation tended to replace concerns with quantifiable results. Thus, the field of information theory had begun by using the old hardware paradigm of transportation of data from point to point. Since electronic information is simultaneously everywhere, the transportation theory yields its relevance to the awareness of transformation of software.

The Western world is very hung up on the problem of visual versus acoustic space, which is delineated in some detail in the next chapter. Western man seems unable to let go of the visual, even though he is floundering in the acoustic world. Gestalt psychology had taken a step away from visual space with its figure-ground paradigm. However, and this is critical to an understanding of tetradic analysis, most psychologists still assume that both figure and ground are visual components in visual situations. In fact, they form an iconic or tactile relationship, defined by the resonant interval between them. That is, there is no continuity or connection in the figure-ground relationship. Instead, there is an interface of a transforming kind. A later chapter explains further this metaphorical positioning.

The degree of confusion that exists in many fields of study with regard to the visual and the acoustic is apparent in Ferdinand de Saussure's *Course in General Linguistics* with his division of language and speech (*la langue* and *la parole*). For Saussure, language is a total and inclusive world of simultaneous structures (that is, right-brain and acoustic), whereas speech, which is sequential, is a relatively superficial and left-brain form, being visual. With these divisions of language and speech, Saussure associated the diachronic and the synchronic:

> But to indicate more clearly the opposition and crossing of the two orders of phenomena that relate to the same object, I prefer to speak of *synchronic* and *diachronic* linguistics. Everything that relates to the static side of our science is *synchronic;* everything that has to do with evolution is *diachronic*. Similarly, *synchrony* and *diachrony* designate respectively a language-state and an evolutionary phase.

Inner Duality and the History of Linguistics

> The first thing that strikes us when we study the facts of language is that their succession in time does not exist insofar as the speaker is concerned. He is confronted with a state. That is why the linguist who wishes to understand a state must discard all knowledge of everything that produced it and ignore diachrony. He can enter the mind of speakers only by completely suppressing the past. The intervention of history can only falsify his judgment.[10]

It probably would have done nothing to clarify these divisions if Saussure had said that the synchronic concerns the acoustic world of the inclusive, the simultaneous, and the unchanging. Even now, the futility of referring to the visual as opposed to acoustic space, resides in the fact that Western man still equates all space with the visual, just as in the eighteenth century all gases were considered variants of air, or pollutions thereof. When an anthropologist such as E. R. Leach turns to expound the thought of Lévi-Strauss, he says:

Lévi-Strauss is distinguished among the intellectuals of his own country, as the leading exponent of "Structuralism," a word which has come to be used as if it denoted a whole new philosophy of life on the analogy of "Marxism"; or "Existentialism." What is this "Structuralism" all about?[11]

When Leach comes to examine the matter, he remarks:

Two features in Lévi-Strauss' position seem crucial. First, he holds that the study of history diachronically and the study of anthropology cross-culturally but synchronically are two alternative ways of doing the same kind of things.[12]

What emerges at once from Leach's approach to Lévi-Strauss is the fact that Leach does not know that the diachronic is visual (left hemisphere) in structure, and that the synchronic is acoustic (right hemisphere) in structure. Having fallen off the rails completely at that early point in his tour of Lévi-Strauss, he not surprisingly fails to relate to Lévi-Strauss in any way whatever. A great deal of what emerges is ignorance of the character of the diachronic and the synchronic, including the fact that these categories, as used by linguists and anthropologists alike, are not understood as presenting the structural clash between the visual and the acoustic.

Elsewhere Leach takes a look behind the work of Lévi-Strauss and discovers:

This, in itself, is no new idea. A much older generation of anthropologists, notably Adolf Bastian (1826–1905) in Germany and Frazer in England held that because all men belong to one species there must be psychological universals (*Elementargedanken*) which should manifest themselves in the occurrence of similar customs among people who had reached the same stages of evolutionary development all over the world. Frazer and his contemporaries assiduously compiled immense catalogs of similar customs which were designed to exhibit this evolutionary principle. This is *not* what the structuralists are up to.[13]

The advantage to this passage is that it reveals another set of hang-ups; namely, that the archetypal and transcendental position, where it concerns psychological universals, is itself based on the use of the paradigm of visual structure to the detriment of acoustic structure. When Coleridge said that all men are born either

Platonists or Aristotelians, he was tending to say that all men tend to be either acoustic or visual, in their sensory bias or preference. But now that this bias has divided the culture of the entire Western world in the electric age, it is no longer a matter of personal temperament or preference, but concerns the very fate of the intelligible, as such. When Leach says "this is *not* what the structuralists are up to," he is also declaring his own unawareness of the difference between the visual and acoustic structures, two totally different forms of cognition. He proceeds to relate the work of Roman Jakobson to Lévi-Strauss and also to Noam Chomsky:

> The influence of Jakobson's style of phonemic analysis on the work of Lévi-Strauss has been very marked; it is therefore relevant that although certain aspects of Jakobson's work have lately been subjected to criticism, Noam Chomsky specifically recognizes the fundamental importance of Jakobson's main theory of distinctive-feature analysis (which reappears in Lévi-Strauss' *Structuralism*) is now rejected by many leading linguists.[14]

The inability of Leach to grasp the different structure of the visual and the acoustic is matched by the similar inability of Jakobson, Lévi-Strauss, and Chomsky, all of whom are unwittingly committed to the structures of visual space with its continuities and homogeneities, rather than to the resonant interfaces of acoustic space. In spite of the failure to recognize the antithetic nature of the visual and the acoustic, those who feel attracted toward structuralism tend to strive to discover inclusive interrelationships in the situations they study.

Visually biased (left hemisphere) people, accustomed to the abstract study of figures minus their ground, are commonly upset by any sudden intrusion of the forgotten or hidden or subliminal ground:

> The human biocomputer is constantly being programmed, continually, simply and naturally, below the levels of its awareness, by the surrounding environment. We noticed that some subjects were quite upset with these effects, which were beyond their immediate control. They would not accept the fact that their brain was reading a word and registering the meaning of that word below their levels of awareness. No matter how hard they tried they could not read the word unless they put their visual axis directly on the word, thus spoiling the experiment. To avoid such

effects of course, we had an observer looking at their eyes and any cases in which they let their eyes move were discounted. This kind of upset was easily corrected by continuing the demonstrations. As the person got used to such results and accepted them, he no longer became upset by the unconscious operations of his biocomputer.[15]

As a cultural mediator, it is the role of the artist to keep the community in conscious relation to the changing and hidden ground of its preferred objectives. Anais Nin writes of D. H. Lawrence:

> . . . Lawrence's characters, whether in poetry, allegory, or prophecy, are actors who speak with the very accents of our emotions; and, before we are aware, our feelings become identified and involved with theirs. Some have recoiled from such an awakening, often unpleasant; many have dreaded having to acknowledge this power of their physical sensations, as well as to face in plain words, the real meaning of their fantasies.

> Lawrence was reviled for going so far. *There are always those who fear for that integral kernel in themselves, for that divine integrity which can be preserved by ignorance* (before psychology) *or by religion* (before and after psychology) *or by the cessation of thought* (by the modern paroxysm of activity).[16]

The task of confronting contemporary man is to live with the hidden ground of his activities as familiarly as our predecessors lived with the figure-minus-ground. In his *Propaganda,* Jacques Ellul explains that the basic condition of the shaping of populations is done not by programs for various media, but by the media themselves, and by the very language we take for granted:

> Direct propaganda, aimed at modifying opinions and attitudes, must be preceded by propaganda that is sociological in character, slow, general, seeking to create a climate, an atmosphere of favorable preliminary attitudes.[17]

After this preparation of the ground, the whole cultural ground itself must be mobilized; not messages but the new configuration of the ground constitutes propaganda:

> Propaganda must be total. The propagandist must utilize all of the technical means at his disposal—the press, radio, TV, movies, posters, meetings, door-to-door canvassing.[18]

That is, the media themselves, and the whole cultural ground, are forms of language. The transforming power of language is recognized by contemporary phenomenology and linguistics as well:

> Further, the usurpation of language does not merely involve the social degradation of words, nor the abuse of the listener's confidence. More profoundly, language inserts itself into the self-consciousness of each man as a screen that distorts him in his own eyes. The intimate being of man is in fact confused, indistinct and multiple. Language intervenes as a power destined to expropriate us from ourselves in order to bring us into line with those around, in order to model us to the common measure of all. It defines and perfects us, it terminates and determines us. The control of consciousness it exercises makes it the accomplice of having, in its monolithic poverty, as opposed to the plurality of being. To the degree that we are forced to resort to language we renounce our interior life because language imposes the discipline of exteriority. The use of speech is thus one of the essential causes of the unhappy conscience, all the more essential because we cannot be without it. It is this which Bruce Parain has strongly emphasized: "At every moment, each consciousness destroys a little bit of the vocabulary it has received and against which it cannot fail to revolt, because it is not its; but immediately it recreates another vocabulary in which it once again disappears."[19]

That is why the human condition seems to the preceding writer a condition of generalized revolt and suicide. The degree to which language as ground biases awareness was very vivid in the experience of the blind Jacques Lusseyran. In his autobiography, *And There Was Light,* he provides an excellent structural or equilibrium approach of his own. This book is an account of the reordering of all of his sensory life as the result of a violent childhood accident in which he lost his sight. This loss of sight greatly enhanced the activity of his other senses and led to the development (or retrieval) of an inner sight as well. Altogether, he became aware that in the sighted world in which he lived, there were a great many assumptions about perception which needed questioning:

> When I came upon the myth of objectivity in certain modern thinkers, it made me angry. So there was only one world for these people, the same for everyone. And all the other worlds were to

be counted as illusions left over from the past. Or why not call them by their name—hallucinations? I had learned to my cost how wrong they were.

From my own experience I knew very well that it was enough to take from a man a memory here, an association there, to deprive him of hearing or sight, for the world to undergo immediate transformation, and for another world, entirely different, but entirely coherent, to be born. Another world? Not really. The same world, rather, but seen from another angle, and counted in entirely new measures. When this happened all the hierarchies they called objective were turned upside down, scattered to the four winds, not even theories but like whims.[20]

In presenting the perceptual patterns of the tetrad form, the object is to draw attention to situations that are still in process, situations that are structuring new perception and shaping new environments, even while they are restructuring old ones, so that it might be said that structures of media dynamics are inseparable from performance. The effort is always to draw attention to the laws of composition as well as to the factors of regulation and interplay.

In *The Study of Human Communication,* Nan Lin stated: "The ultimate goal of science is to explain by means of a set of theories, events that are observed."[21] The McLuhan tetrad is designed to do just that; and is not based on a theory or a set of concepts, but, rather, it relies on observation and experience and percepts. While empirical, it provides a basis for prediction (e.g., of what will be retrieved or what reversals of form will occur).

We have indicated earlier that all human artifacts are extensions of man, outerings or utterings of the human body or psyche, private or corporate. As utterances they are speech, translations from one form into another form, be it hardware or software: metaphors. Of course, all words, in every language, are metaphors. Structurally speaking, a metaphor is a technique of presenting one situation in terms of another situation. That is to say it is a technique of awareness, of perception (right hemisphere) not of concepts (left hemisphere). As two situations are involved, there are two sets of figure-ground relations in apposition, though the grounds may or may not be stated. All metaphors have four components in analogical ratio. Thus, "cats are the crabgrass of life" presents "cats are to (my) life as crabgrass is to an otherwise beautiful lawn." Or, "she sailed into the room" presents "her motion entering the room" in terms of a ship's swift (perhaps graceful) motion under sail. To say that metaphor has four terms which are discontinuous, yet in ratio to one another, is to say that the basic mode of metaphor is resonance and interval—the audile-tactile. (See Chapter 1.) This discontinuity was pointed out by Aristotle in *De Anima*:

> It follows that the soul is analogous to the hand; for as the hand is a tool of tools, so the mind is the form of forms and sense the form of sensible things.[22]

Apropos the four-part structure which relates to all human artifacts (verbal and nonverbal), their existence is certainly not deliberate or intentional. Rather, they are a testimony to the fact that the mind of man is structurally active in all human artifacts and hypotheses. Whether these appositional ratios are also present in the structure of the natural world raises an entirely separate question. It is perhaps relevant to point out that the Greeks made no entelechies or observations of the effects of man-made technology, but only for what they considered the objects of the natural world.

The usual approach to metaphor is purely verbal rather than operational or structural, that is, in left-hemisphere terms of the figures only, minus their grounds. Thus, metaphor is discussed as a form of sort-crossing or of category mistake or of misnaming. For example:

> However appropriate in one sense a good metaphor may be, in another sense there is something inappropriate about it. This inappropriateness results from the use of a sign in a sense different from the usual, which use I shall call "sort-crossing." Such sort-crossing is the first defining feature of a metaphor and, according to Aristotle, its genus: "Metaphor (*meta-phora*) consists in giving the thing a name that belongs to something else; the transference (*epi-phora*) being either from genus to species . . . *or* on the grounds of analogy." (*Rhetoric* 1457b)[23]

Elsewhere in the *Rhetoric*, Aristotle betrays his left-hemisphere visual bias in his confusion of metaphor and simile:

The simile also is a metaphor; the difference is but slight. When Homer says of Achilles (cf. *Iliad* 20.164),

"He sprang at the foe as a lion."

that is a simile. When he says of him, "The lion sprang at them," it is a metaphor; here, since both are courageous, the poet has transferred the name of "lion" to Achilles. Similes . . . are to be employed in the same way as metaphors, from which they differ in the point just mentioned.[24]

Aristotle regards both figures as concepts and as propositional, whereas the character of the metaphor is discontinuous, abrupt, and appositional. His approach is descriptive rather than structural or perceptual.

Paul Ricoeur has devoted *The Rule of Metaphor* to an examination and discussion of recent approaches to metaphor from various disciplines, including linguistics, semantics, the philosophy of language, literary criticism, and aesthetics. In discovering the Aristotelian notion of metaphor as alien usage—the substitution theory—he makes a revealing slip regarding his own assumptions about words.

Now the fact that the metaphorical term is borrowed from an alien domain does not imply that it substitutes for an ordinary word which one could have found in the same place. Nevertheless it seems that Aristotle himself was confused on this point and thus provided grounds for the modern critiques of the rhetorical theory of metaphor. The metaphorical word takes the place of the non-metaphorical word that one could have used (on condition that it exists), so that it is doubly alien, as a present but borrowed word and as a substitute for an absent word.[25]

However, all words are metaphors. The non-metaphorical word is a feature of primitive tribal thought (about words) only. The native hunter or Eskimo says "Of course 'stone' *is* stone, otherwise how could I know stone?" If some words, e.g., names, are non-metaphorical, then metaphorical expression is impossible as there can be no balance of ratio or proportion. All that remained would be synecdoche or simile or metonymy. But language always preserves the play or figure-ground relation between experience (or perception) and its replay in expression. It is this same left-hemisphere approach to the right-hemisphere (appositional) properties of language that causes Ricoeur to relegate all technology to the domain of *logos* rather than to that of mythos.

At the heart of Ricoeur's approach to metaphor is the transportation theory of communication. It is, he remarks, "the relationship between the embryonic classification of Aristotle and the concept of transposition, which constitutes the unity of meaning of the genus 'metaphor.' "[26] Ricoeur continues:

> Two facts should be noted. First transposition operates between logical poles. Metaphor occurs in an order already constituted in terms of genus and species, and in a game whose relations—rules—subordination, co-ordination, proportionality or equality of relationships—are already given. Second, metaphor consists in a violation of this order and this game. In giving to a genus the name of a species, to the fourth term of the proportional relationship the name of the second term, and vice versa, one simultaneously recognizes and transgresses the logical structure of language (1457b 12–20). The *anti* discussed earlier, applies not just to the substitution of one word for another, but also to the jumbling of classification in cases that do not have to do only with making up for lexical poverty. Aristotle himself did not exploit this idea of a categorical transgression, which some modern authors compare to Gilbert Ryle's concept of "category mistake." Doubtless this was because he was more interested, within the perspective of his *Poetics,* in the semantic gain attached to the transference of names than in the logical cost of the operation. The reverse side of the process, however, is at least as interesting to describe as the obverse.[27]

In other words, at this stage of his analysis, Aristotle was more involved with either-or than with both-and. Ricoeur is trying to hold the discussion of metaphor in terms of matching instead of the making process, in terms of a special kind of logic and dialectic instead of poesis, of concept (descriptive) instead of percept. To do so, it is necessary to ignore ground and interpolate a dialectic of polar figures, to reduce proportion to equalities (which robs them of resonance), and to preserve a logical structure of language. Thus, he speaks of (Aristotelian) analogy, "which we have seen, is analyzed as an identity or similarity of two relations," and of "the logical moment of proportionality."

Ricoeur's main problem, and that of most contemporary rhetorical criticism, is due to the confusion that arises from not dealing with something on its own terms. Throughout his discussion, Ricoeur leans on Aristotle's distinction of metaphor as part of rhetoric on the one hand, and as part of mimesis on the other. His essential point is contained in Aristotle's statement, "to metamorphize well implies an intuitive perception of the similarity in dissimilars"—that is, unlike things (both-and). Full explication of the unresolved and unquestioned assumptions in Ricoeur's thoughts, and indeed all modern examination of metaphor would require an extensive history of the medieval trivium—grammar, dialectic, and rhetoric. As yet, no such history exists, though portions of it are available, e.g., in the works of Jaeger, Howells, Ong, Lubac, and Marron, to mention a few. However, the explications of some of these authors suffer from not accounting for the interdependence and interaction of the "three roads."

Intense rivalry characterized the medieval trivium from the outset. As dialecticians, Plato's and Aristotle's accounts of rhetoric are extremely biased. For this reason, although Aristotle's may be the oldest extant systematic treatise on rhetoric, it is not to be relied upon for a rhetorician's awareness of the art. Rather it presents a dialectical understanding: a modern parallel might be a handbook by Heidegger on advertising techniques. Ricoeur himself is located somewhere between dialectic and grammar.

The trivium, the ancient reformulation of the arts or sciences of the *logos,* was born of the phonetic alphabet. The effect of the phonetic alphabet on the Greek psyche and culture was catastrophic. Mimesis gave way to individualized detachment, and the integral resonating oral *logos* was broken into multiple fragments, each bearing some one or another of its properties. For over a century, in the early Greek era, there was a great number of these systems, including poets, exegetes, philosophers, rhetors, et cetera, and combinations of these. But it was the fifth-century Stoics who formulated the essential tripartite relationship. The Stoics developed a three-fold *logos* that served as a pattern for the later trivium, although the trivium itself was not formally recognized as the basis of education and science for some time. The pre-alphabetic *logos* was retrieved in two ways: it informed the patristic doctrine of the Logos, and it was recapitulated in the overlapping structures of the three-fold Stoic *logos.*

Briefly, the relation between the Stoic system and the trivium is as follows. Their *logos hendia thetos* is an inner word, or, the abstract word prior to (or minus) speech: this idea structurally adumbrates dialectic (logic and philosophy) with its left-hemisphere emphasis on abstraction (figure-minus-ground) and absolutes, and on correct thought form (sequence), irrespective of content or audience. Their *logos prophorikos* is the uttered word and corresponds to rhetoric as the science of transforming audiences with speech. Their *logos spermatikos* is the uttered *logos* as seeds embedded in things animate and inanimate which structures and informs them and provides the formal principles of their being and growth (becoming). This *logos* is the root of grammar (the Greek word for which is "literature" in the Latinate rendering), with its twin concerns of etymology and multiple-level exegesis, the ground search for structure and roots. All of the sciences (e.g., the later quadrivium of music, arithmetic, geometry, and astronomy) were, structurally, subdivisions of grammar, as forms of exegesis of (the book of) nature; to which they are returning today. Ancient rhetoric and grammar, then, are principally right-brain activities: a dialectical rendering of either (such as Plato's or Aristotle's), quite aside from partisanship, would be a metaphor for or a translation of the original.[28]

From the beginning, the trivium was beset by rivalry between the brain hemispheres, later known in Swift's time as the wars of the Ancients and the Moderns, with grammar and rhetoric usually holding control of the trivium against the rival claims of the dialecticians. Cicero, following the rhetorician Isocrates (a contemporary of Plato) and Quintilian, established the basic pattern for civilized education in the West, whether of prince or poet (reaffirmed by St. Augustine four centuries later), as the alignment of encyclopedic wisdom and eloquence. That is, the conjunction of grammar—the tradition of learned exegesis and commentary—and rhetoric provided a balance of the hemispheres.

For these practitioners, tradition had the same right-hemisphere figure-ground resonance and simultaneity that T. S. Eliot (a modern grammarian of ancient ilk) had proposed in his essays on poesy. For more than fifteen centuries, most of our Western history—the Ciceronian program—itself a retrieval of the old Greek liberal educational system (the *enkyklios paideia*), as the trivium was a retrieval of the oral *logos,* was the basis of liberal education and humanism.

With print, via Gutenberg, the visual stress of the alphabet gained new ascendancy.[29] Spearheaded by the French dialectician Peter Ramus, a new battle of the Ancients (rhetoricians and grammarians) and Moderns was waged, and the dialectic method took over from tradition. Since that time grammar and rhetoric have been cast in a dialectic or left-hemisphere mold, along with all our arts and sciences. It is only with the return to acoustic space in our world, to right-hemisphere multi-sensory forms of awareness, that the tables begin to turn once more.

Hence, the perceptual patterns of the tetrad form belong properly to grammar, not to philosophy in its present rhetorical guise. Our concern in this book is etymology and exegesis. The etymology of all human technologies is to be found in the body itself: they are, as it were, prosthetic devices, mutations, metaphors of the body or its parts. (This fact you might discover for yourself if you were a quadriplegic, and could control your environment only by blowing into a plastic reed, the pulsations of which were interpreted by a bedside computer; and, which, in turn, raised and lowered the bed, called for the nurse, turned book pages, and put lights on and off, etc.)

The tetrad is exegesis on four levels, showing the *logos*-structure (not mythos) of each artifact; its four parts as metaphor or word. This is to place for the first time the whole study of technology and artifacts on a humanistic and linguistic basis, one which is "valueful" rather than valueless. Nevertheless, the true metaphoric character of the tetrad cannot be illuminated, for the first-time student, without an understanding of how phonetic literacy separated the perception of space into the arenas of the visual and the acoustic, our chief concern in the next chapter.

Visual and Acoustic Space

Visual space is a side effect of the uniform, continuous, and fragmented character of the phonetic alphabet, originated by the Phoenicians and enlarged by the Greeks. Some neurologists and sociologists have claimed that hierarchical reasoning is a sensory preference of the left hemisphere of the brain; and, audile-tactile space a sensory preference of the right brain, the dwelling place of primitive man's intuition of myth. Physiology of the eye may have prompted beginnings of linear logic.

Cash money and the compass, leading technologies of the fifteenth century, illustrate early figure-ground transformations of visual space archetypes to the acoustic, from the tangible to the intangible, from hardware dominance to software dominance—analogous to the present role of the computer. Current shift from visual space to acoustic space technologies in society is accelerating.

While in elementary school, Jacques Lusseyran was accidentally blinded. He found himself in another world of collision and pressure points. No longer could he pick his way through the ordinary "neutral" world of reflected light. It was the same environment that we are all born into but now it came to him demanding exploration:

> Sounds had the same individuality as light. They were neither inside or outside, but were passing through me. They gave me my bearings in space and put me in touch with things. It was not like signals that they functioned but like replies . . .

But most surprising of all was the discovery that sounds never came from one point in space and never retreated into themselves. There was the sound, its echo, and another sound into which the first sound melted and to which it had given birth, altogether an endless procession of sounds . . .

Blindness works like dope, a fact we have to reckon with. I don't believe there is a blind man alive who has not felt the danger of intoxication. Like drugs, blindness heightens certain sensations, giving sudden and often disturbing sharpness to the senses of hearing and touch. But, most of all, like a drug, it develops inner as against outer experience, and sometimes to excess. . . .[1]

We, who live in the world of reflected light, in visual space, may also be said to be in a state of hypnosis. Ever since the collapse of the oral tradition in early Greece, before the age of Parmenides, Western civilization has been mesmerized by a picture of the universe as a limited container in which all things are arranged according to the vanishing point, in linear geometric order. The intensity of this conception is such that it actually leads to the abnormal suppression of hearing and touch in some individuals. (We like to call them "bookworms.") Most of the information we rely upon comes through our eyes; our technology is arranged to heighten that effect. Such is the power of Euclidean or visual space that we can't live with a circle unless we square it.[2]

But this was not always the expected order of things. For hundreds of thousands of years, mankind lived without a straight line in nature. Objects in this world resonated with each other. For the caveman, the mountain Greek, the Indian hunter (indeed, even for the latter-day Manchu Chinese), the world was multicentered and reverberating. It was gyroscopic. Life was like being inside a sphere, 360 degrees without margins; swimming underwater; or balancing on a bicycle. Tribal life was, and still is, conducted like a three-dimensional chess game; not with pyramidal priorities. The order of ancient or prehistoric time was circular, not progressive. Acoustic imagination dwelt in the realm of ebb and flow, the *logos*. For one day to repeat itself at sunrise was an overwhelming boon. As this world began to fill itself out for the early primitive, the mind's ear gradually dominated the mind's eye. Speech, before the age of Plato, was the glorious depository of memory.

Acoustic space is a dwelling place for anyone who has not been conquered by the one-at-a-time, uniform ethos of the alphabet. It exists in the Third World and vast areas of the Middle East, Russia, and the South Pacific. It is the India to which Gandhi returned after twenty years in South Africa, bringing with him the knowledge that Western man's penchant for fragmentation would be his undoing. There are no boundaries to sound. We hear from all directions at once. But the balance between inner and outer experience can be precise. If our eardrums were tuned any higher we would hear molecules colliding in the air or the roaring rush of our own blood. Sound comes to us from above, below, and the sides. As Lusseyran says, it passes through us and is rarely limited by the density of physical objects. Most natural materials act as a tuning fork. The human baby cannot move out into the environment until sound teaches depth—which the child adapts to the demands of Euclidean or visual space later on.

Each of these modalities is a sensory preference of the culture. For the society that accepts it, that modality, whether acoustic or visual, is the foundation on which it recognizes its own perception of sanity. But we wish to advance an idea that you, the reader, won't in all probability, initially accept. And that is for several thousand years, at least, man's sensorium, or his seat of perceptive balance, has been out of plumb.

The term *senus communis* in Cicero's time meant that all the senses, such as seeing, hearing, tasting, smelling, and touch, were translated equally into each other. It was the Latin definition of man in a healthy natural state, when physical and psychic energy were constant and distributed in a balanced way to all sense areas.[3] In such a condition it is rather difficult to hallucinate. In any cultural arrangement, trouble always occurs when only one sense is subjected to a barrage of energy and receives more stimulus than all the others. For modern Western man that would be the visual state.

As psychologists understand sense ratios, overstimulation and understimulation can cause thought and feeling to separate. Sleeping may be regarded as a dimming down of one or two sensory inputs. Hypnosis, on the other hand, is a steady assault on one sense, like a tribal drumbeat. Modern torturers in Chile break down prisoners by putting them in cells where everything—walls, furniture, utensils, window covers—is painted white. In Vietnam, Communist interrogators discovered (as police interrogators everywhere) that unexpected beatings and random electric shocks create sharp peaks of floating anxiety and subsequently a ready uncritical conviction.

Without being aware of it, North Americans have created the same kind of violence for themselves. Western man thinks with only one part of his brain and starves the rest of it. By neglecting ear culture, which is too diffuse for the categorical hierarchies of the left side of the brain, he has locked himself into a position where only linear conceptualization is acceptable.

Euclid and Newton fixed Western man's body in rigid space and oriented him toward the horizon.[4] As neurosurgeon Joseph Bogen puts it, the linear sequential mode of the left hemisphere underlies language and analytical thought. The right hemisphere of the brain, which is principally concerned with pattern recognition of an artistic and holistic quality, grasps the relationship between diverse parts readily and is not bound up with a rigid sequence of deductions. The intellectual legacy of Euclid and Newton therefore is a substitution of perspective for qualitative thinking, which is always composed of multi-sensual elements.

Everything in life after the Greeks was reduced to the uniform and the homogenous, Swift's island of Laputa. Thought had to have a beginning, a middle, and an end. No thesis was acceptable unless all ideas were interconnected to project an e-x-t-e-n-d-e-d point of view, which is the interior structure of the essay, we might add.

If you think of every human sense as creating its own space, then the eye creates a space where there can only be one thing at a time. The eye acts as a machine—like a camera. Light focused on the back of the eye ensures that two objects will not occupy the same place at the same time. The mind teaches the eye to see an object right side up, on a plane and in perspective space. As children, when perspective (or the vanishing point) arrives—when we learn to focus an inch or two in front of the page—we learn to read and write. The phonetic alphabet gives us a point of view since it promotes the illusion of removing oneself from the object.

It would almost seem that the very physiology of the eye promotes the idea that everything is in sequence—that is, in its proper place, at the proper time, and in linear relationship. The kind of mentality that prompted Shakespeare's King Lear to divide his kingdom among his daughters, to abstract himself from the medieval perception that England was contained in himself is more modern than tribal. What we are saying is that the human eye appears to be the father of linear logic. Its very nature encourages reasoning by exclusion: something is either in that space or it isn't.

The constraints of Western logic are tied to our sense of sequential relationships—logic made visual. The middle ground, however accounted for initially, is eventually excluded. It is either-or. If your culture nurtures you to favor the eye, your brain has difficulty giving equal weight to any other sense bias. You are trapped by visual only assumptions. For centuries, the Japanese, unlike Westerners, have treasured the pictorial space between objects in a painting, the *ma;* and have viewed such space as more dominant than all objects portrayed. Like the yin/yang complementarity of wave/particle in atomic physics.

Anyone who has been involved in gestalt, or studied primitive societies—once he or she gets over the impulse to measure these societies with Western templates—is aware that either-or is not the only possibility. Both-and can also exist. People who have not been exposed to the phonetic alphabet, that is, the "uncivilized," can easily entertain two diametric possibilities at once. Edmund Carpenter pointed out to us that the Inuits, or the Eskimos, cannot visualize in two dimensions. If they are asked to draw the animals they hunt on a flat surface, the result—to our eyes—is often grotesque. But ask them to draw the same figure on, let us say, the rounded surface of a walrus tusk, and the etched drawing will take on three-dimensional life as you roll the tusk in your fingers. Siegfried Giedion tells us:

> E. S. Carpenter kindly sent me the reproduction of a reindeer knife-handle (Royal Ontario Museum, Toronto) which depicts a caribou in two characteristic positions: one on guard, the other grazing. If the handle is turned through ninety degrees, the grazing animal becomes upright and watchful. . . .[5]

So here we have a clue to the mentality of the pre-literate, that world of oral tradition that we eventually left behind about the end of the Hellenic period. It is the mentality of the multitude, or as Yeats put it: everything happening at once, in a state of constant flux. For the genuinely tribal man there is no causality, nothing occurring in a straight line. He turns aside from the habit of construing things chronologically—not because he can't, but as Edmund Carpenter says, because he doesn't want to.

Carpenter advises us that the Trobriander Islanders only recognize now, the eternal present. Bronislaw Malinowski and Dorothy Lee, who studied these people, discovered that they disdained the concept of *why*. European man to them was hung up on the idea of setting priorities, of making past and future distinctions. "To the Trobriander, events do not fall of themselves into a pattern of cause and effect as they do for us. We in our culture automatically see and seek relationships, not essence. We express relationship mainly in terms of cause or purpose. . . ."[6] The Trobriander is only interested in experiencing the current essence of a person or object. He is interested in his yams, his stone knife, his boat, as those objects are today. There is no such thing as a "new" or an "old" boat, a blooming yam or a decayed one. There is no past or future, only the essence of being that exists now. The Trobriander, like the Inuit, directly experiences a sense of timelessness, so he is never bothered by such questions as "who created the creator." The English language, in fact most Western languages, suggests through its tense structure that reality can only be contained in the concept of a past, a present, and a future—which rather incongruously implies that man is capable, like a god, of standing outside the time continuum. The hubris of Western man might very well lie in the priority-setting propensity for quantitative reasoning.

In a world of simultaneous comprehension involving the right hemisphere of the brain, chronology does not dominate. For the image patterns of the right brain, all events have the quality of equal time. And when dealing with the megatrends of history, and human behavior in general, it is better to analyze the broad implications of movement patterns rather than a particular event. Nowhere is this caveat better illustrated than in the development and distribution of cash money, which has been one of the major definers of Western cultural relationships since Lydia floated her coins on the commercial sea of Asia Minor—what we call "process pattern recognition."

Process pattern recognition is another term for tetradic analysis. When we began to describe the action of the tetrad to a beginner, we discovered that the artifacts of cash money and the compass were useful teaching aids; everyone seemed to know something about them. They were particularly useful in describing the four-fold action of the tetrad in figure-ground (for example, $A/D = C/B$ and $B/D = C/A$). The four-fold pattern clearly demonstrates that the true tetrad had two grounds and two figures in a balanced ratio to each other, which tends to highlight the nature of the reversal stage. Here again we want to remind the reader that a tetrad is a device which does not detail technological change but only the distinctive features of innovation in human terms. It is more like an ideograph than a treatise.

The Chinese were the first to find out that coin money and barter brought a kind of vortex of power as these economic processes abraded one another. While Marco Polo was in Kubla Khan's court, he encountered the use of paper money and was advised that long before, in the ninth century B.C., iron coins had been circulated. If that was so, the Great Khan and his successors should have experienced all the ills associated with such a public exchange, including forgery and inflation.

Barter, however exquisitely the Orientals may have practiced it, has one basic drawback: it takes a lot of time to find the person who might want your goods or services and to agree on items of equal value. Harold Innis says that in any early marketing economy—such as the Northwest trading post—population traffic forces a coin-like standard of value to arise, over-riding barter, perhaps in the same way that Babylonian clay tablets became a medium of due bills and receipts.

Whether coinage is wampum, tobacco leaves, or beaver skins, the rate of transaction, which some economists call the velocity of money, rises as more people come to market. Barter and coin money (keyed to bullion weight) pace each other for a while, but then there comes a time when coin money takes over and eclipses barter. At the high point of obsolescence, conspicuous consumption displays itself most readily, such as Louis the Fourteenth's gold service or Trimalchio's extravagant mock Roman funerals.

The appearance of potlatch, or a conspicuous show of wealth, triggers a change in the trading climate. And almost simultaneously a reversal takes place where the *image* of wealth becomes more important than the actual ability to repay. In medieval Europe, deposits at the goldsmith's shop, and the consequent issuance of deposit notes, performed this function. Tangible money is transmuted into the intangible, or creditworthiness, the essence of paper currency and fiat checks.

This transfer from the tangible to the intangible and back again has been played and replayed a number of times in the history of the West. After the fall of Rome, in A.D. 476, coinage rapidly declined, and for almost a thousand years the halting Near East and European economies subsisted on renewed barter and small bullion exchange. Continental trade routes collapsed and the cities drew in on themselves once again. Constantinople lost its place as a trade mediator between East and West; but after 1453, the northern city-states of Italy pushed themselves forward as high traffic markets and began to issue coinage. Despite periodic fluctuations, the overall pattern of money usage, which covers at least seven thousand years of record, could be expressed tetradically (see Fig. 3.1).

Coinage and the compass acted as the chief trade emblems of the fifteenth century. The Italians got the compass from the Arabs, who, in turn, picked it up from the Chinese, perhaps centuries earlier. Metal money might be looked upon as an archetypal kind of hardware because ultimately all things in Western culture are invariably reduced to cash units of value. The compass by the late Renaissance had almost as powerful an impact, only for different reasons. Money is one of man's outerings from his body, an externalization of his need to exchange and store property. The compass is another outering, an externalization of his sense of direction on this planet. What is important about these artifacts is that by the fifteenth century, in the age of exploration, they had begun to intensify in man's consciousness.

The Chinese had compass bowls in the fifth century B.C. By the fifteenth century A.D., China was sending vast treasure fleets, led by an admiral called Zheng He, toward Africa and Arabia. The largest fleet consisted of sixty ships and twenty-seven thousand men. Zheng He commanded an oceangoing junk of 140 meters, the largest ship of its time, complete with a liquid azimuth compass.[7]

Cash Money

A. Speeds
transactions

D. Reverses
into credit
or non-money

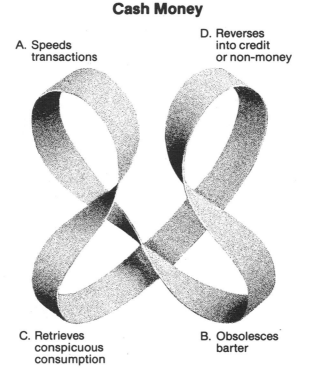

C. Retrieves
conspicuous
consumption

B. Obsolesces
barter

Fig. 3.1.

The original compass might be considered a primitive form of circuitry. It persuaded the user to think of the world as a spherical plenum. So, for Cabot and Columbus, the strength of the compass is that it made a square earth into a round one.

A tetrad of the compass shows that man's environment is not only changed by the artifacts man invents, but that he also reassesses his world every time he meets a strange artifact. The artifacts make him look backward as well as forward. Any mariner today, viewing a copy of any early liquid or pivot compass—indeed, imagining the Yellow Emperor's south-pointing carriage of some two thousand years ago—would instantly recognize a prototype of the inertial guidance system (see Fig. 3.2).

Any artifact, whether hardware (compass, cash, chairs, com-

Compass

A. Enhances range
and accuracy of
global navigation

D. Reverses into
a cosmic
environment

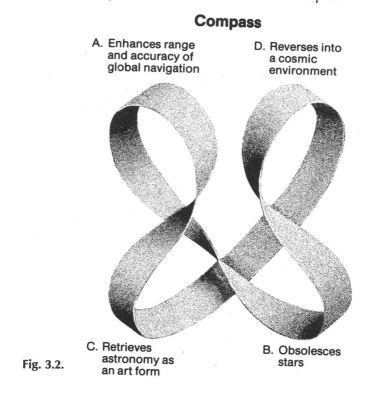

C. Retrieves
astronomy as
an art form

B. Obsolesces
stars

Fig. 3.2.

puter) or software (art styles, poetry, philosophy, scientific theories) when subjected to intense speedup will eventually reach a state of information overload in which the process pattern of the artifact as an aspect of eternity becomes clear. It is at that stage that we know that particular extension of our senses for what it really is in human, not necessarily technological, terms. Most Western methods of analysis usually portray the effects of physical outering in positive terms, as a single efficient cause, that is, how can we use it? The tetrad serves us by pointing up the holistic effects as well, by giving a formal cause view of all processes working together, that is, by revealing the interplay of all tetradic elements as ground.

The tetrad is configurational—a viewing of all sense processes abrading one another. Being configurational, it may be seen in its multi-dimensional character as acoustic. The acoustic properties reveal what the element of visual space cannot, and vice versa. Yet both these views of the spatial mode are evoked by environmental influences.

To summarize, visual space structure is an artifact of Western civilization created by Greek phonetic literacy. It is a space perceived by the eyes when separated or abstracted from all other senses. As a construct of the mind, it is continuous, which is to say that it is infinite, divisible, extensible, and featureless—what the early Greek geometers referred to as *physis*. It is also connected (abstract figures with fixed boundaries, linked logically and sequentially but having no visible grounds), homogeneous (uniform everywhere), and static (qualitatively unchangeable). It is like the "mind's eye" or visual imagination which dominates the thinking of literate Western people, some of whom demand ocular proof for existence itself.

Acoustic space structure is the natural space of nature-in-the-raw inhabited by non-literate people. It is like the "mind's ear" or acoustic imagination that dominates the thinking of pre-literate and post-literate humans alike (rock video has as much acoustic power as a Watusi mating dance). It is both discontinuous and nonhomogeneous. Its resonant and interpenetrating processes are simultaneously related with centers everywhere and boundaries nowhere. Like music, as communications engineer Barrington Nevitt puts it, acoustic space requires neither proof nor explanation but is made manifest through its cultural content. Acoustic and visual space structures may be seen as incommensurable, like history and eternity, yet, at the same time, as complementary, like art and science or biculturalism.

Occasionally, certain persons in history have been in the right place and time to be truly bicultural. When we say bicultural we mean the fortune to have a foot placed, as it were, in both visual and acoustic space, like Hemingway in his Cuban village hideaway or Toqueville in America. Marco Polo was such a one. The Phoenicians, the earliest cultural brokers between East and West, having brought a cuneiform method of accounting to the Egyptians and the phonetic alphabet to the Greeks, were likewise blessed.

The phonetic alphabet underlies all of Western linguistic development.[8] By the time it had gone through the Greeks and Romans and reasserted itself in the print literature of the Renaissance, Western sense ratios had been firmly altered. The Greeks gave a new birth to the alphabet as a mode of representation having neither visual nor semantic meaning. Egyptian ideographs, for instance, were directly related to particular sensuous sounds and actions, with unique graphic signs. On the other hand, the matrix of the Greek alphabet could be used to translate alien languages back and forth without changing the form and number (twenty-four) of the original alphabetic characters. It became the first means of translation of knowledge from one culture to another. The reader in the process became separated from the original speaker and the particular sensuous event. The oral tradition of the early Greek dramatists, of the pre-Socratics, and Sophocles, gave way very gradually to the written Pan-European tradition and set the emotional and intellectual posture of the West in concrete, as it were. We were "liberated" forever from the resonating magic of the tribal word and the web of kinship.

The history of the Western world since the time of Aristotle has been a story of increasing linguistic specialism produced by the flat, uniform, homogeneous presentation of print. Orality wound down slowly. The scribal (or manuscript) culture of the Middle Ages was inherently oral/aural in character. Manuscripts were meant to be read aloud. Church chantry schools were set up to ensure oral fidelity. The Gutenberg technology siphoned off the aural-tactile quality of the Ancients, systemized language, and established heretofore unknown standards for pronunciation and meaning. Before typography there was no such thing as bad grammar.

After the public began to accept the book on a mass basis in the fourteenth and fifteenth centuries—and on a scale where literacy mattered—all knowledge that could not be so classified was tucked away into the new "unconscious" of the folk tale and the myth, there to be resurrected later as the Romantic Reaction.

But since World War I and the advent of those technical wavesurfers Marconi and Edison, the rumbles of aural-tactility, the power of the spoken word, have been heard. James Joyce in *Finnegans Wake,* celebrated the tearing apart of the ethos of print by radio, film (television), and recording. He could easily see that Goebbels and his radio loudspeakers were a new tribal echo.

And you may be sure that emerging mediums such as the satellite, the computer, the data base, teletext-videotext, and the international multi-carrier corporations, such as ITT, GTE, and AT&T, will intensify the attack on the printed word as the "sole" container of the public mentality, without being aware of it of course. By the twenty-first century, most printed matter will have been transferred to something like an ideographic microfiche as only part of a number of data sources available in acoustic and visual modes.

This new interplay between word and image can be understood if we realize that our skulls really contain two brains straining to be psychically united. The tetrad, which is a psychic aid in understanding that relationship, will be examined later in this book as a diagram of the bifurcated mind.

East Meets West in
the Hemispheres

Ordering facility of the left brain is quantitative (diachronic): reading, writing, naming within a parameter of significant hierarchy. The right brain is the area of the qualitative (synchronic), wellspring of the spatial-tactile, the musical and the acoustic. When these hemispheric functions are in true balance, which is rare, "comprehensive awareness" is the result. True consciousness has always had a diachronic and synchronic character. The corpus callosum in the brain potentially promotes a healthy interchange of labeling and imaging between the hemispheres. The Western world, especially Europe and the Americas, emphasizes left-brain thinking over right-brain cognition. We should be aware that cultures exist, like the Inuit, where the reverse is true.

In our desire to illumine the differences between visual and acoustic space, we have undoubtedly given a false impression: and that is that the normal brain, in its everyday functioning, cannot reconcile the apparently contradictory perceptions of both sides of the mind. There is, we know from experience, a "unified field" of the mind. We learn a piano piece by working through the lines of individual musical notes, but the music does not spring to life for us until we sense its overall harmonic structure.

There are countless examples in art, science, and technology of visual and acoustic space working together to fabricate, more or less, a melded and consistent idea of the outside world. One of the most influential, at least for linguists, is Ferdinand de Saussure's hypothesis of language development. He wanted to find a way of expressing a method for separating *la parole* (individual speech) from *la langue* (the established language).

Saussure contended that a language existed only in terms of how it was spoken by a large group of people in a definitive geographic area. People spoke as they felt and reinforced each other's use of the mother tongue by peer pressure.[1] Yet, inexorably, over a period of years, the language would change as new connotations appeared. Hence, it had a propensity to intrinsically form itself through use, a synchronic structure. And, at the same time, it also responded to a movement through time, its diachronic nature. The diachronic always took an apposite, or axial, relationship to the synchronic (Fig. 4.1).

The diachronic line (CD) in the drawing can be viewed as the action of visual space and the synchronic line (AB) as that of the acoustic. Saussure was really involved in explicating a right-brain phenomenon but his idea winds up looking like a geometric construct; it is almost as if we cannot express an exercise in pattern recognition unless we give it a linear orientation. This is our way of saying that Western culture habitually forces us to approach the right hemisphere of our cortex through the agency of the left hemisphere.

The key to our future development as a species will depend on how well we understand the relationship between the left side and right side of our associative cortex and the utility of those millions of nerve fibers connecting the two sides called the corpus callosum.[2] We must teach ourselves to abandon the tendency to view the environment in a hierarchical and totally connective way, to center ourselves instead in the arena of interplay between the two modes of perception and analysis, which is comprehensive awareness. Linus Pauling shook the foundations of classical physics by reminding his fellow scientists that nothing in the material universe connects. The same thing can be said of the mind; all its elements interface.

On the outside, in a grossly anatomical way (Fig. 4.2), the brain appears to be what it seems, walnutlike and symmetrical, covered all over with convoluted fissures designed to give more tissue area. But underneath, within that three pounds of whitish

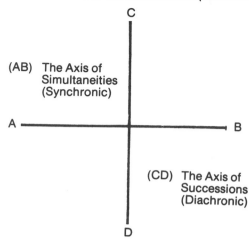

Simultaneities/Successions (Saussure)

Fig. 4.1.

mush, is a seething electrochemical mass which has the power to function asymmetrically.[3] Normally, of course, we do not know it. If we decide to go for a jog, the left hemisphere, through the corpus callosum, sends a signal to the right hemisphere to move both hips synchronously. As we write, a capacity largely controlled by the left posterior lobe, the right hemisphere guides the curlicues of the West's Palmer Method. Millions of neural interfaces keep us coordinated. Although, if we might take a slight detour of thought, the fact that neurons never actually connect, or touch, should be of immense interest to neurophysiologists. When an electrical impulse reaches the tip of a neuron's tail, or axon, it discharges a chemical called a neurotransmitter. This chemical message diffuses across a gap called a synapse to receptors located in the next cell, triggering yet another electrical charge that courses through another axon until the message reaches millions of other neurons. The brain, it would appear, is a mosaic that resonates in its "discrete" parts.[4]

The left side of the upper brain has a very specialist role. It is largely concerned with linguistic matters, the ability to order, to quantify, to label. The right side of the neo-cortex is best in spa-

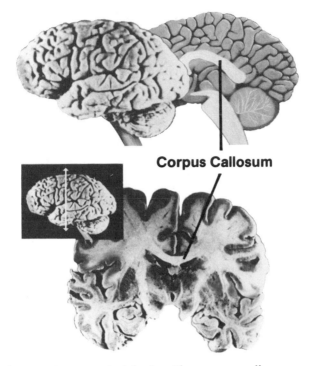

Corpus Callosum

Fig. 4.2. Asymmetrical brain: The corpus callosum as a thick band of nerve fibers joins the left and right brain and coordinates audile/spatial construction and nonverbal ideation (right hemisphere) with calculation, speech, writing, and general linguistic abilities (left hemisphere). Hearing, left and right visual fields, as well as handedness are associated on a cross-over basis. For most people the left brain is dominant; the right brain ("minor") seems primarily involved with audile-tactile areas and pattern recognition. After Restak.

tial tasks, the sense of the multi-dimensional. The field of vision in each eye is divided between the left and right brain. Josephine Semmes tells us that the left hemisphere prefers units of neural information which can be said to be "similar" and focal, whereas the right brain area favors unintegrated units of data.[5] The back, or posterior, lobes of the upper brain, which deal with specific touch sensations and spatial information, interact vigorously with the frontal lobes, which tend to abstractly play with the constraints of time and the ability to plan—the world of now linked to the world of future.

And then there is the lower brain, the various levels of earlier tissue which evolved over millions of years: the spinal cord and the brain stem that attend to such basics as heartbeat and respiration; the R-Complex, reptilian seat of aggression, ritual, territoriality, social hierarchy; the limbic system, cradle of our emotions. Paul MacLean says that the upper brain lives in an uneasy peace with the lower and mid-brain and what we may actually be talking about are three separately interfacing cognitive systems.

However fascinating such a theory may be, we should focus on the relationship between the cortical hemispheres which in our view is the projection of consciousness, consciousness being the sum interaction between one's self and the outside world.[6] The hemispheres manifest their very nature by the way in which they perceive and analyze the environment. Carl Sagan says this ability is the unique mark of the primate.

The areas of Broca and Wernicke in the left cortical hemisphere centralize our capacities for speech, hearing, and writing and thus mediate our expression of comprehension and language. The left hemisphere is the seat of hierarchies and categories, of the linear, the mathematical, and the sequential. The ordering sense of the left brain is quantitative (the diachronic): reading, writing, naming within a perception of significant order. We are not surprised, therefore, when neuroscientists place the guidance of complex sensory-motor skills, like typing or adjusting a micrometer, in that part of the cortex.

The left hemisphere has mastery over the right side of the body; the right hemisphere dominates the left side. Joseph Bogen, in his seminal treatise, "The Other Side of the Brain," isn't quite sure what the right brain does, but he thinks it could very well be the birthplace of creativity.[7] Certainly it is the field of the qualitative (the synchronic): the spatial-tactile, the musical, and the acoustic. It has been called the mute part of the brain because its language abilities are minimal, but there is considerable evidence that the right side analyzes by configuration and by metaphor. It does not think in sequence but rather in terms of seizing the relationship between unlike parts of the environment. The right brain perceives the essence of an object through shape and "feel" rather than naming classification.

The right and left brain hemispheres actually pursue two different thinking and analytic processes; that is, two different ways of processing information. This state of affairs was described brilliantly by Robert J. Trotter, writing on an investigation of the patterns of behavior "among the Inuit or Eskimo people of Baffin Island in northeastern Canada."[8] Headed by anthropologist Solomon Katz of the University Museum of the University of Pennsylvania, the investigation "dealt specifically with one of the most fascinating and fastest growing areas of brain research, cerebral asymmetry or hemispheric dominance."

He called attention to the fact that those living in the arctic wastes have a well-documented ability to find their way out of difficult wilderness situations as well as to navigate over vast areas of unmarked territory of snow and ice. The land itself appears to have given them a special visuo-spatial ability, related to an intricately developed right hemisphere. But of more importance, in their life style and art objects there seems to be a well-defined cooperation between right and left brain:

> Among the Inuit carvers (all of whom were right-handed), the left hand cradles the work, moves it into the new positions and feels its progress while the right hand precisely carves the details and holds the various carving tools. Even when a tool could be placed down, the left hand carried out the repositioning of the stone in space. Also . . . there was a striking preponderance of holding the stone in the left visual field. . . .

> These observations suggest hemispheric symmetry or at least a high degree of cooperation between the hemispheres. Katz finds an almost perfect relationship between the right hand doing the detailed analytical kinds of activities and the left hand doing all the spatial and *touch* activities.[9]

Similarly, what the Baffin Island investigators discovered among the Inuit people was a language reflecting "a high degree of spatial, right hemisphere orientation. Linguistic studies rate it as being the most synthetic (integrative) of languages. American English is at the other end of the same scale, and is rated as the most

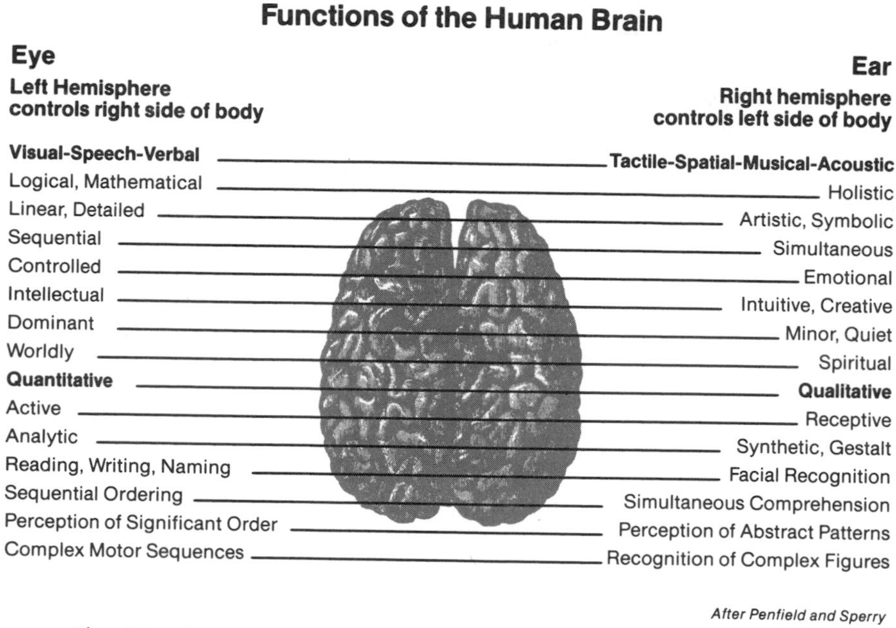

Functions of the Human Brain

Eye	Ear
Left Hemisphere **controls right side of body**	**Right hemisphere** **controls left side of body**

Visual-Speech-Verbal	**Tactile-Spatial-Musical-Acoustic**
Logical, Mathematical	Holistic
Linear, Detailed	Artistic, Symbolic
Sequential	Simultaneous
Controlled	Emotional
Intellectual	Intuitive, Creative
Dominant	Minor, Quiet
Worldly	Spiritual
Quantitative	**Qualitative**
Active	Receptive
Analytic	Synthetic, Gestalt
Reading, Writing, Naming	Facial Recognition
Sequential Ordering	Simultaneous Comprehension
Perception of Significant Order	Perception of Abstract Patterns
Complex Motor Sequences	Recognition of Complex Figures

After Penfield and Sperry

Fig. 4.3. Chart-diagram of brain functions, after R. J. Trotter's artwork in *Science News;* see adaptation above adopted at Centre for Culture and Technology, 1978.

analytic (left-hemisphere)." Inuit sculptures, lithographs, and tapestries are "without apparent linear or three dimensional analytic orientation." The culture is audile-tactile to a high degree.

Hence, if you are an Inuit, your mental faculties have a sensory preference toward the right hemisphere and a world of sensuous touch and echoing ritual form. The right leads the left. If you are from the West, and particularly a well-urbanized person, your mind will tend to favor the left and interplay with a somewhat subservient right. With that in view, one can read Trotter's chart-diagram of the left and right brain (Fig. 4.3).[10]

Because the strongest feature of the left hemisphere in the West is linearity and sequentiality, there are good reasons for calling it the visual (quantitative) side of the brain; its most conspicuous role seems to be that of labeling. Because the dominant features of the right hemisphere are the simultaneous, holistic, and synthetic, there are good reasons for identifying it as the acoustic (qualitative) side—palpable imagery not bound by time.[11]

Visual space as elucidated in Euclidean geometry has the basic characteristics of lineality, connectedness, homogeneousness, and stasis. These characteristics are not found in any of the other senses. On the other hand, acoustic space has the basic character of a sphere whose focus or center is simultaneously everywhere and whose margin is nowhere.[12] An accident to the left hemisphere might limit speech or produce aphasia. But "damage exclusively to the right hemisphere does not usually disrupt linguistic abilities but can lower performance in spatial tasks, simple musical abilities, recognition of familiar objects and faces and bodily self-awareness."

This is another way of saying that visual and acoustic space are always present in any human situation, even if Western civilization has—through the agency of the alphabet—tamped down our awareness of the acoustic. The latter is the invisible counter-environment that forms the background against which the civilization of the written word is seen. Christian von Ehrenfels, the originator of Gestaltist structuralism, clearly showed that configurations (gestalten) exist only because of our tendency to see figure against ground, to prefer geometrically perfect forms against irregular shapes.[13]

That Trotter, in "The Other Hemisphere" essay, selects a Third World or non-literate society for observation and illustration points also to the fact that societies that have not developed the use of the phonetic alphabet tend to adopt the same Third World posture. The Third World is mainly oral-aural, even when it cultivates some nonphonetic form of writing such as Sanskrit. On the other hand, First World countries tend to be visual (left hemisphere) even when most of their population is declining into a semi-literate state. Such is the case today when the visual culture of industrial societies has been greatly influenced in an acoustic direction by the environment of electronic technologies.

The dominance of the left or the right hemisphere, in terms of controlling the principal problem-solving of the brain at any one time, is largely dependent on environmental factors; so that in the West the lineality of the left hemisphere is supported—for example—by a complex service environment of roads and transportation and logical or rationalistic activities in legal administration. The dominance of the right hemisphere, on the other hand, depends upon an environment of a simultaneous resonating character, a point that we shall develop in some detail in Chapter 5 when discussing the Oriental view of space.[14] Such dominance is normal in oral societies, keeping in mind that the oral is closely related to the tactile, particularly among pre-literates. Today, our universal environment of simultaneous electronic flow, of constantly interchanging information, favors the sensory preference of the right hemisphere. The First World is aligning itself, however gradually, with the Third World.

Western man depends on his capacity for conceptualizing proportional space to confirm measurable fact. The Inuit, as indeed do all peoples of an Oriental disposition, find truth is given not by "seeing is believing," but through oral tradition, mysticism, intuition, all cognition—in other words, not simply by observation and measurement of physical phenomena. To them, the ocularly visible apparition is not nearly so common as the auditory one. *Hearer* would be a better title than *seer* for their holy men.

Plato and Angelism

Orientals have a capacity for instant readjustment to all psychic and social conditions, which is related to seeing life as a multi-sensory equilibrium with no ordering priorities. The ground of experience is constantly tuned. All ground; no figures—the acoustic mode. Westerners are hung up on a fixed point of view where ground is fragmented, engendering a desire for hierarchy—all figures; no ground. Robotism is the ability to be equally empathetic in many areas at once. Angelism is being chained to a fixed point of view, without ground.

Plato's early struggles were centered on destroying the Oriental bias of Greek tribalism; he emphasized left-brain cognition over right, or angelism.

Plato saw the intrusion of the mimetic and resonant characters from the mental theater of the right hemisphere as a kind of "psychic poison," or threat to his left-hemisphere academy with its goal of an individualized polis:

> You threw yourselves into the situation of Achilles, you identified with his grief or his anger. You yourself became Achilles and so did the reciter to whom you listened. Thirty years later you could automatically quote what Achilles had said or what the poet had said about him. Such enormous powers of poetic memorization could be purchased only at the cost of total loss of objectivity. Plato's target was indeed an educational procedure and a whole way of life. . . . This kind of drama, this way of reliving expe-

rience in memory instead of analyzing and understanding it, is for him "the enemy."

The empathy of the right hemisphere is incompatible with the detachment of the left hemisphere:

> That is why the poetic state of mind is for Plato the arch-enemy and it is easy to see why he considered this enemy so formidable. He is entering the lists against centuries of habituation in rhythmic memorized experience. He asks of men that instead they should examine this experience and rearrange it, that they should think about what they say, instead of just saying it. And they should separate themselves from it instead of identifying with it; they themselves should become the "subject" who stands apart from the "object" and reconsiders it and analyzes it and evaluates it, instead of just "imitating" it.[1]

The alphabet created a lineal and visual environment of services and experiences (everything from architecture and highways to representational art) which contributed to the ascendancy or dominance of the left, or lineal, hemisphere. This conjecture is consistent with the findings of the Russian neurophysiologist A. R. Luria, who found that the area of the brain which controls linear sequencing, and hence mathematics and scientific thinking, is located in the pre-frontal region of the left hemisphere:

> The mental process for writing a word entails still another specialization: Putting the letters in the proper sequence to form the word. Lashley discovered many years ago that sequential analysis involved a zone of the brain different from that employed for spatial analysis. In the course of our extensive studies we have located the region responsible for sequential analysis in the anterior regions of the left hemisphere.[2]

Luria's results show that the expression "linear thinking" is not merely a figure of speech, but a mode of activity which is peculiar to the anterior regions of the left hemisphere of the brain. His results also indicate that the use of the alphabet, with its emphasis on linear sequence, stimulates mastery of this area of the brain in cultural patterns. Luria's observations provide an understanding of how the written alphabet, with its lineal structure, was able to create the conditions conducive to the development of Western science, technology, and rationality.

The alphabet separated and isolated visual space from the many other kinds of sensory space involved in the senses of smell, touch, kinesthesia, and acoustics. Abstract visual space is lineal, homogeneous, connected, and static. When, however, neurophysiologists assign a vague "spatial" property to the right hemisphere, they are referring to the simultaneous and discontinuous properties of audile-tactile and multiple other spaces of the sensorium.

The Euclidean space of analytic geometry is a concept of the left hemisphere of the brain, while the multi-dimensional spaces of the holistic sensorium are percepts of the right hemisphere of the brain. Where the phonetic alphabet comes into play, the visual faculty tends to separate from the other senses, making possible the perception of abstract Euclidean (visual) space, which represents an extreme dislocation or dissociation from the other senses. The history of the rise of Euclidean geometry, especially in the time of ancient Alexandria, offers a direct parallel with the rise of phonetic literacy. And, phonetic literacy, in turn, is co-existent and co-extensive with the rise of rational logic.

Parmenides is the first visual (quantitative) philosopher, and he succeeds the pre-Socratics who were right-hemisphere, acoustic (qualitative) philosophers. The phonetic alphabet obsolesced the oral culture of Greece, as is explained in *Preface to Plato* by Eric Havelock:

> Both pre-Socratics and Sophists then, by the close of the fifth century before Christ, if the *Apology* does indeed reproduce the idiom of that period, were accepted by public opinion as representative of the intellectualist movement. If they were called "philosophisers," it was not for their doctrines as such, but for the kind of vocabulary and syntax which they used and the unfamiliar psychic energies that they represented. Sophists, pre-Socratics and Socrates had one fatal characteristic in common; they were trying to discover and to practise abstract thinking.[3]

The power of the phonetic alphabet to translate other languages into itself, to act as a porous matrix of information flow, is matched by its power to invade right-hemisphere (oral) cultures. In the ordinary way, these tribal, right-hemisphere cultures are holistic and entire and resistant to penetration by other pre-literate cultures. But the specialist qualities of the left-hemisphere phonetic alphabet have long offered the only instrumental means of invading and taking over oral societies.

"Propaganda cannot succeed where people have no trace of Western culture." These words of Jacques Ellul in *Propaganda* draw attention to one of the crucial features of Western history. The Christian church, dedicated to propaganda and propagation, adopted Greco/Roman phonetic literacy from its earliest days. The perpetuation of Greco/Roman literacy and civilization became inseparable from Christian missionary and educational activity. Paradoxically, people are not only unable to receive but are unable to retain doctrinal teaching without a minimum of phonetic or Western culture. Here is the observation of Ellul on this matter:

> In addition to a certain living standard another condition must be met: if man is to be successfully propagandized, he needs at least a minimum of culture. Propaganda cannot succeed where people have no trace of Western culture. We are not speaking here of intelligence; some primitive tribes are surely intelligent, but have an intelligence foreign to our concepts and customs. A base is needed—for example, education; a man who cannot read will escape propaganda, as will a man who is not interested in reading. People used to think that learning to read evidenced human progress; they still celebrate the decline of illiteracy as a great victory; they condemn countries with a large proportion of illiterates; they think that reading is a road to freedom. All this is debatable, for the important thing is not to be able to read, but to understand what one reads, to reflect on and judge what one reads. Outside of that, reading has no meaning (and even destroys certain automatic qualities of memory and observation). But to talk about critical faculties and discernment is to talk about something far above primary education and to consider a very small minority. The vast majority of people, perhaps 90 percent, know how to read, but they do not exercise their intelligence beyond this. They attribute authority and eminent value to the printed word, or, conversely, reject it altogether. As these people do not possess enough knowledge to reflect and discern, they believe—or disbelieve—*in toto* what they read. And as such people, moreover, will select the easiest, not the hardest, reading matter, they are precisely on the level at which the printed word can seize and convince them without opposition. They are perfectly adapted to propaganda.[4]

Phonetic literacy in Athens and Greece was an intensely disruptive force, as explained by Philip Slater in *The Glory of Hera,* and by Karl Popper in *The Open Society and Its Enemies.* Slater is concerned with the breakup of Greek family life and the rise of the new democratic and competitive individualism. There was a pronounced reaction against all the new qualities of mind and spirit released by the impact of literacy. Popper asks: "How can we explain the fact that outstanding Athenians like Thucydides stood on the side of reaction against these new developments?" He notes that:

> while many of the ambitious young nobles became active although not always reliable members of the democratic party, some of the most thoughtful and gifted resisted its attraction . . . the open society was already in existence . . . and had, in practice, begun to develop new values, new equalitarian standards of life. . . .[5]

The big tribal leaders of Athens tried hard to resist what in effect was a violent transition from the holistic (acoustic) right-hemisphere institutions of their oral society to the fragmented and scientific bias of the visual revolution evoked by the literate activation of the left hemisphere. Plato was by no means happy about the effects of literacy and the rise of aggressive commercial interests:

> Although the "patriotic" movement was partly the expression of the longing to return to more stable forms of life, to religion, to decency, law and order, it was itself morally rotten. Its ancient faith was lost, and was largely replaced by a hypocritical and even cynical exploitation of religious sentiments. Nihilism, as painted by Plato in the portraits of Callicles and Thrasymachus, could be found if anywhere among the young "patriotic" aristocrats who, if given the opportunity, became leaders of the democratic party.

The same literacy, which destroyed the traditional institutions of Athens, created an abstract rationalism inseparable from the new dominance of the left hemisphere:

> But at this time, in the same generation to which Thucydides belonged, there arose a new faith in reason, freedom, and the brotherhood of all men—the new faith, and, as I believe, the only possible faith, of an open society.[6]

Karl Popper, who had no awareness of the role of literacy in the revolution from Greek tribalism to Greek individualism, or in separating the individual from the group, sums up:

> . . . we can say that the political and spiritual revolution which had begun with the breakdown of Greek tribalism reached its climax in the fifth century, with the outbreak of the Peloponnesian War. It had developed into a violent class war, and at the same time, into a war between the two leading cities of Greece.[7]

This individualism may be what made Greco/Roman institutions attractive to Christianity, since Christian revelation stresses the private responsibility of all individuals in its doctrine of the resurrection. The phonetic alphabet creates the only environmental services and institutions which foster the dominance of the left hemisphere.

The dominance of the left hemisphere (analytic and quantitative)—and by dominant we mean the ability of the left brain to lead the right brain in the context of Western culture—entails the submission or suppression of the right hemisphere; and so, for example, our intelligence tests exist only for measuring left-hemisphere achievement and take no cognizance of the existence of the (qualitative) right hemisphere. The present electronic age, in its inescapable confrontation with simultaneity, presents the first serious threat to the 2500-year dominance of the left hemisphere.

There are a variety of factors which can give salience or mastery either to the right (simultaneous and acoustic) hemisphere of the brain, or to the left (lineal and visual) hemisphere. No matter how extreme the dominance of either hemisphere in a particular culture, there is always some degree of interplay between the hemispheres, thanks to the corpus callosum and the anterior and hippocampal commissures—that part of the neural cable network which bridges the hemispheres.

Even the Chinese with their extreme cultivation of the right hemisphere—which invests every aspect of their lives, their language, their writing, with artistic delicacy—exert much left-hemisphere bias and quality in their practicality and concern with moral wisdom. However, their stress falls heavily on what Heisenberg calls the resonant interval or touch. It is a matter of the experience of time and space. A Westerner, for example, arranges flowers *in* space;

the Chinese and the Japanese harmonize the space *between* the flowers. The importance of this discontinuous space becomes clear in the following passage from *The Chinese Eye* by Chiang Yee:

> Indeed, the use of space is one of the Chinese painter's most coveted secrets, one of the first thoughts in his head when he begins to plan his composition. Almost every space in our pictures has a significance: the onlooker may fill them up with his own imagined scenery or with feeling merely.

> There was a Chinese poet of the Sung dynasty, Yeh Ch'ing Ch'en, who wrote the sorrows of a parting and described the scene as follows:

> Of the three parts Spring
>   scene, two are sadness,
> And the other part, is nothing
>   but wind and rain.[8]

The Chinese, in other words, allow the right hemisphere to direct the left; they use the eye as an ear, creating the seemingly paradoxical situation that Tony Schwartz notes in *The Responsive Chord* apropos the TV image: "In watching television, our eyes function like our ears."[9]

The experience of the right brain leading the left is by no means foreign to us. Herbert Krugman performed brain-wave studies, comparing the response of subjects to print and television. One subject was reading a book as the TV came on. As soon as she looked up, her brain waves slowed significantly. Within thirty seconds, she was in a predominantly alpha state—relaxed, passive, unfocused. Her brain wave response to three different types of TV content was basically the same, even though she told Krugman she liked one, disliked another, and was bored by a third. As a result of a series of such experiments, Krugman argues that this essential alpha state is characteristic of how people respond to TV—any TV. He remarked:

> . . . the ability of respondents to show high right brain response to even familiar logos, their right brain response to stories even before the idea content has been added to them, the predominantly right brain response to TV, and even perhaps to what we call print advertising—all suggests that in contrast to teaching, the unique power of the electronic media is to *shape* the content

of people's imagery, and in that particular way determine their behavior and their views.[10]

Krugman's investigations were, he admitted, initially undertaken to disprove the "medium is the message." His quantitative results point to a massive and subliminal erosion of our culture through right-hemisphere indoctrination by TV. In a wider sense all electronic media as a new configuration or ground give salience only to the right hemisphere. There is no way of quantifying the right hemisphere, which emphasizes inner and qualitative aspects of experience.

The current spate of dyslexia and other learning disability difficulties—some 90 percent of the victims are male—may well be a direct result of TV and other electronic media pressuring us into returning to the right hemisphere.[11] Dyslexia is an inability to adopt a single, fixed point of view with respect to all letters and words. Conversely, it consists of approaching letters and words from many points of view simultaneously (i.e., right-brain fashion), without the assumption that any one view is solely correct.

As near-point-vision problems build, so will problems with a left-hemisphere alphabetic form. Is there a direct relationship between soaring delinquency rates in the U.S. and Canada and reading disability? The cubists, as artists and the "antennae of the race," detected the shift some seventy years ago and explored the emerging grammar of the right-brain sensory modality. If literacy is to survive in the West, our writing system will soon have to be recast in a mold congenial to right-hemisphere sensibilities and satisfactions. It may be necessary, for example, to make a transition from stick-and-ball penmanship to italics, from the Palmer Method to a shortened form of calligraphy—in brief, everyman a cubist.

We have long been accustomed to using the interval between the wheel and the axle as an example not only of touch, but also of play. Without play, without that figure-ground interval, there is neither wheel nor axle. The space between the wheel and the axle, which defines both, is "where the action is"; and this space is both audile and tactile. The Chinese, as we have said, use the interval between things as a primary means of getting in touch with situations.

Nothing could be more expressive than this interval of the properties of the right hemisphere in contrast to the left, for to the left brain, the interval is a space which must be logically connected, filled and bridged. Such is the dictate of lineality and visual order in contrast to the resonating interval or gap of the simultaneous world of the right hemisphere. In *The Book of Tea,* Okakura Kakuzo explains the Japanese attitude to social relationships as a "constant readjustment to our surroundings." This is the extreme contrast to the Western or visual point of view which assumes a fixed position from which to examine each situation and to assert one's preference. Right-hemisphere culture has no place for the private individual, just as left-hemisphere society regards tribal groups as sinister and threatening (e.g., the "Yellow Peril"). World War II soldiers and sailors used to refer to natives in the South Pacific ("the boondocks") as "gooks," to express their apartness, an attitude continued in Vietnam.

> Suzuki, the great authority on Zen Buddhism, describes muga as "ecstasy" with no sense of *I am doing it,* "effortlessness." The "observing self" is eliminated; a man "loses himself," that is, he ceases to be a spectator of his acts. Suzuki says: "With the awakening of consciousness, the will is split into two . . . actor and observer. Conflict is inevitable, for the actor (minus self) wants to be free from the limitations of the observer-self. Therefore, in enlightenment, the disciple discovers that there is no observer-self, 'no soul entity as an unknown or knowable quantity.' Nothing remains but the goal-and-the-act that accomplishes it."[12]

The right-hemisphere culture has a great affinity for the simultaneity of the age of electronic information, as Okakura Kakuzo explains:

> The present is the moving infinity, the legitimate sphere of the relative. Relativity seeks adjustment: adjustment is art. The art of life lies in a constant readjustment to our surroundings.[13]

The right-hemisphere culture naturally seeks to tune or reconfigure intervals rather than to connect situations and relationships:

> The Taoists claimed that the comedy of life could be made more interesting if everyone would preserve the unities. To keep the proportion of things and give place to others without losing one's own position was the secret of success in the mundane drama. We must know the whole play in order to properly act our parts; the

conception of totality must never be lost in that of the individual. This Laotse illustrates by his favourite metaphor of the Vacuum. He claimed that only in the vacuum lay the truly essential. The reality of a room, for instance, was to be found in vacant space enclosed by the roof and walls, not in the roof and walls themselves. The usefulness of a water pitcher dwelt in the emptiness where water might be put, not in the form of the pitcher or of the material of which it was made. Vacuum is all potent because it is all containing. In vacuum alone motion becomes possible.[14]

One can see immediately how the concept of zero must have struck the West when it was first introduced through Africa to Europe in the Middle Ages. This Oriental idea suggested the possibility of play in mathematics and science, the absence of which tended to cripple the computations of the Romans and the Greeks. Kakuzo adds: "In Jiu-Jitsu one seeks to draw out and exhaust the enemy's strength by non-resistance, vacuum, while conserving one's own strength for victory in the final struggle." In Western art, we admire the power of statement and the bounding line in design, whereas the right-hemisphere culture gives play to the opposite principle; instead of statement, the stress is on "the value of suggestion":

> In art the importance of the same principle is illustrated by the value of suggestion. In leaving something unsaid the beholder is given a chance to complete the idea and thus a great masterpiece irresistibly rivets your attention until you seem to become actually a part of it. A vacuum is there for you to enter and fill up to the full measure of your aesthetic emotion.[15]

This is of the same order as the pre-literate Greek technique of mimesis, discussed earlier.

Yet, some of our most perceptive Western writers, particularly in the early part of the century, were so captured by the rigidity of a fixed point of view, of objectivity, that they failed to give equal time to the reconciliation of opposites. They did not understand that oral cultures regard robotism as a desirable norm. In contrast, Wyndham Lewis, in *Men Without Art,* maintains that the role of the "civilized" artist is to prevent our becoming adjusted, since to individualized Western society the "well-adjusted" man is an impercipient automaton.

The term "robotism" therefore, as we use it, does not mean the mechanically rigid behavior of "Rossum's Universal Robots," as Karel Capek used the word in his 1938 play. Rather robotism in this context means the suppression of the conscious "observer-self," or conscience, so as to remove all fear and circumspection, all encumbrances to ideal performance. Such a man, as Suzuki says, "becomes as the dead, who have passed beyond the necessity of taking thought about the proper course of action. The dead are no longer returning *on;* they are free. Therefore to say 'I will live as one already dead' means a supreme release from conflict."[16]

The Japanese use "living as one already dead" to mean that one lives on a plane of expertness. It is the extinction of the left-hemisphere detached and objective self. If the result resembles detachment, it is from pushing the right hemisphere to a state of total enlargement or enhancement, to the point of reversal (chiasmus) of apparent characteristics. As Ruth Benedict remarks, "it is used in common everyday exhortation. To encourage a boy who is worrying about his final examinations from middle school, a man will say, 'take them as one already dead and pass them easily.' To encourage someone who is undertaking an important business deal, a friend will say, 'Be as one already dead.' When a man goes through a great soul crisis and cannot see his way ahead, he quite commonly emerges with the resolve to live 'as one already dead.' " She continues:

> It points up vividly the difference between Western and Eastern psychology that when we speak of a conscienceless American we mean a man who no longer feels the sense of sin which should accompany wrongdoing, but that when a Japanese uses an equivalent phrase he means a man who is no longer tense and hindered. The American means a bad man; the Japanese means a good man, a trained man, a man able to use his abilities to the utmost. He means a man who can perform the most difficult and devoted deeds of unselfishness. The great American sanction for good behavior is guilt; a man who because of a calloused conscience can no longer feel this has become antisocial. The Japanese diagram the problem differently.

> According to their philosophy, man in his inmost soul is good. If his impulse can be directly embodied in his deed, he acts virtuously and easily. Therefore he undergoes, in "expertness," self-

training to eliminate the self-censorship of shame (*haji*). Only then is his "sixth sense" free of hindrance. It is his supreme release from self-consciousness and conflict.[17]

The paradox today is that the ground of the latest Western technologies is electronic and simultaneous, and thus is structurally right-hemisphere and "Oriental" and oral in its nature and effects. The reason for this state of affairs is that most Western technologies retain, often quite unnecessarily—like the computer—a mechanical one-thing-at-a-time bias, as a form of nineteenth-century cultural lag. This situation began with the telegraph over a century ago. The overwhelming pattern of procedures in the Western world remains lineal, sequential, and connected in political and legal institutions and also in education and commerce, but not in entertainment or art—a formula for complete chaos.

The ground of the Oriental right-hemisphere world, meantime, is rapidly acquiring some of the hardware connectedness of the left-hemisphere Western world. Before 1940, Japan began to compete in the arena of modern industrialism in textiles and shipbuilding. Today it rivals American competence in the making of semiconductors and small autos. China has recently embarked on a program of mass alphabetic literacy, which will result in her acquiring a completely left-hemisphere cultural bias, plunging the Chinese into a new phase of individualized enterprise and aggression, for which they are already developing, once again, a ground of industrial hardware. As one Canadian observer who had been there in 1980 with a Canadian symphony orchestra says: "The Chinese are in the act of tearing their cultural skin off in order to protect themselves against Russian hardware."

In general, it needs to be noted that left-hemisphere man has very little power to observe or control environments, or to see the patterns of change. The Oriental tradition, on the other hand, reflects a particular attunement to all facets of ground and immediate responsiveness to changes in ground configuration. In the 1970s the Japanese quickly perceived the developing public need for small cars, on a world scale. American car makers were mesmerized by the car as object (figure); the Japanese saw the car as ground, as part of the service environment which encompassed it.

Oral peoples, in Asia or the Third World, are notoriously conservative about new technologies because of their sensitivity to the side effects involved—the new ground they bring into play—and theirs are histories of rejections of innovations. Westerners, in contrast, tend to adopt anything that promises an immediate profit and to ignore all side effects. It is this sensitivity to ground, plus a strong sense of decorum (propriety) and a lack of private identity that enables an Oriental to change behavior instantly from one pattern to another. For example, until August 1945 the *chu* code of loyalty demanded of the Japanese people that they fight to the last man against the enemy. When the emperor changed the requirements of *chu* by broadcasting Japan's capitulation, the Japanese outdid themselves in expressing their cooperation with their victors.

> Occidentals cannot easily credit the ability of the Japanese to swing from one behaviour to another without psychic cost. Such extreme possibilities are not included in our experience. Yet in Japanese life the contradictions, as they seem to us, are as deeply based in their view of life as our uniformities are in ours. It is especially important for Occidentals to recognize that the "circles" into which the Japanese divide life do not include any "circle of evil." This is not to say that the Japanese do not recognize bad behaviour, but they do not see human life as a stage on which forces of good contend with forces of evil. They see existence as a drama which calls for careful balancing of the claims of one "circle" against another and of one course of procedure against another, each circle and each course being in itself good.[18]

Instead of an abstract or objective uniform (visual) code or conduct applicable to all situations (i.e., as figure minus a ground) there is rather a (multi-sensory) equilibrium or balance of properties to be adjusted constantly. The ground must remain tuned. Robotism is instant readjustment.

Angelism on the other hand ensures a rigidity of point of view which is largely a consequence of linear and visual logic. It is best characterized as promoting confrontation and fragmentation, some of the chief elements in the illusion of objectivity. One emphasizes the eye over the ear. The function of robotism is the reverse. As

Lowell Thomas used to say, "On the air, you're everywhere. . . ."
The robotic man is capable of instant adjustment to any social situation without guilt; since he keeps his ear tuned to a collective, a
moral identity which we call the audience. Like the attentive crowd,
an audience is tuned ground.

Hidden Effects

All of man's artifacts, of language, of laws, of ideas and hypotheses, of tools, of clothing and computers, all of these are extensions of the human body. Man cannot trust himself with his own artifacts. The tetrad is needed to reveal any artifact's subliminal effects. Every artifact is an archetype, and the ongoing cultural recombination of old and new artifacts is the engine of all invention and drives the subsequent wide use of invention, which is called innovation.

If you have ever sat in a hot and airless lecture room trying to follow the speaker's line of argument, you have experienced the psychic nature of a figure: it is the momentary area of your mind's attention. As you sit there, you will notice perhaps successively a sudden shift in the air, the radiator knocking, an insect buzzing between the screen and the pane, or the pressure of your legs against the chair. Within the context of all the things that exist in that room, points of awareness (attention) will arise and recede. In a larger sense, nothing has meaning except in relation to the environment, medium, or context that contains it. The type on this page is the figure against the ground of the blank page. The figure of the geometric construct is revealed against the void in which it is imagined. The left hemisphere of the brain is figure against the ground of the right brain in Western culture and the opposite for the Oriental.

In his book *Out of Revolution,* Eugen Rosenstock-Huessy explains how the figure of Western capitalism has persisted in a program of advance by environmental destruction, without any policy of replacement of such (environmental) ground. By contrast, the right-hemisphere man, like the primitive hunter, who has learned to move through nature rather than against it, is always intensely aware of ground and, in fact, prefers ground and the experience of participation in ground to the detached contemplation of figures. Chiang Yee points to the rejection of (visual) matching and representation in Chinese art:

> Verisimilitude is never a first object; it is not the bamboo in the wind that we are representing but all the thought and emotion in the painter's mind at a given instant when he looked upon a bamboo spray and suddenly identified his life with it for a moment.

He further notes:

> . . . we try, in the steps of the Sages, to lose ourselves in Great Nature, to identify ourselves with her. And so in landscapes, in the paintings of flowers and birds, we try not to imitate the form, but to extract the essential feeling of the living object, having first become engulfed in the general life stream.

The Oriental aspires not merely to love and understand a painting itself, but to probe for a meaning far beyond its confines in a world of the spirit. On these right-hemisphere terms, figure painting is a peculiar Western preoccupation that is devoid of satisfaction:

> We have never elevated figure-painting as you have in the West; some of it may have religious significance, but it seldom reaches the depth of thought which landscape attains.[1]

Until the advent of the expressionists and the cubists, art in the West was in thrall to Renaissance perspective and individual portraiture, requiring a detached observer. By tuning in on the new audile-tactile awareness made available these days by our electronic ground, Fritjof Capra found that modern physicists were, unwittingly, retrieving a worldview which is harmonious with ancient Eastern wisdom. His problems in reconciling the two were entirely those of the hemispheres:

> I had gone through a long training in theoretical physics and had done several years of research. At the same time, I had become very interested in Eastern mysticism and had begun to see the parallels to modern physics. I was particularly attracted to the

puzzling aspects of Zen, which reminded me of the puzzles of quantum theory. At first, however, relating the two was a purely intellectual exercise. To overcome the gap between rational, analytical thinking and the meditation experience of mystical truth, was, and still is, very difficult for me.[2]

A. R. Luria's observations provide an understanding of how the written alphabet with its lineal structure was able to create the conditions conducive to the development of the Western mental ethos, especially science, technology, and rationality. Many left-hemisphere stroke patients become aphasic, losing some or all of their ability to speak or to write and, in some cases, also losing the capacity for sustained (sequential) thought. They seem to become "astonied" (fifteenth-century English) or stunned—the experience not unlike being stoned on drugs or alcohol.

In part, this condition may be due to a loss of muscular motor control. But much of it is directly related to the inner-outer split between the hemispheres and to the linearity feature of the left side of the brain. In effect, some stroke victims are unceremoniously dumped into the processes of the other hemisphere which is not proficient in reading, writing, and naming. Left-brain centered speech and writing has to be uttered in a sequence. Just as all forms of sequential activity (as contrasted to configurational or pattern) are functions of the left hemisphere, so too all forms of utterances (and artifacts), whether technological, verbal, or written, are functions of the left hemisphere.

This extends to private identity—uttering of the self as fragmented and abstracted from the group—and to entrepreneurial aggression of all kinds. Conversely, all technologies that emphasize the outer or the abstract or sequentiality in organizing experience, contribute to left-hemisphere dominance in a culture. Harold Innis remarked on the Oriental (right hemisphere) antipathy to sequence and abstraction and precision:

Social time, for example, has been described as qualitatively differentiated according to the beliefs and customs common to a group and as not continuous but as subject to interruption of actual dates. It is influenced by language which constrains and fixes prevalent concepts and modes of thought. It has been argued by Marcel Granet that the Chinese are not equipped to note concepts or to present doctrines discursively. The word does not fix a notion with a definite degree of abstraction or generality but evokes an indefinite complex or particular image. It is completely unsuited to formal precision. Neither time nor space is abstractly conceived: time proceeds by cycles and is round. . . .[3]

Dr. Joseph Bogen, the surgeon who participated in the initial split brain operations with Phillip Vogel, noted, appositely, "what may well be the most important distinction between the left and right hemisphere modes is the extent to which a linear *concept* of time participates in the ordering of thought."[4] It was the dominance of the left hemisphere by means of the civilizing stream of phonetic literacy, linked with the time concept, that enabled Western man to detach himself from participation in his surroundings. His program to conquer nature is but one result of the enormous psychic and cultural energy released by that ground of specialist goals.

It is always the psychic and social grounds, brought into play by each medium or technology, that readjust the balance of the hemispheres and human sensibilities which are in equilibrium with those grounds. As we mentioned before, the experience of youthful Jacques Lusseyran in his blinded state amply illustrates how the shift of any component in the sensorium creates an entirely different world:

> When I came upon the myth of objectivity in certain modern thinkers, it made me angry. So there was only one world for these people, the same for everyone. And all the other worlds were to be counted as illusions left over from the past. Or why not call them by their name—hallucinations? I had learned to my cost how wrong they were.

> From my own experience I knew very well that it was enough to take from a man a memory here, an association there, to deprive him of hearing or sight, for the world to undergo immediate transformation, and for another world, entirely different but entirely coherent, to be born. Another world? Not really. The same world rather, but seen from another angle, and counted in entirely new measures. When this happened, all the hierarchies they called objective were turned upside down, scattered to the four winds, not even like theories, but like whims.[5]

Lusseyran was made particularly aware of the right-hemisphere inner experience afforded by blindness by having lived in an objective left-hemisphere culture. Blindness creates the *seer* much as the ancient world conceived the *seer* as blind.

> Blindness works like dope. . . . But, most of all, like a drug, it develops inner as against outer experience, and sometimes to excess.[6]

In our culture the parallel is the caricature of inner or right-hemisphere awareness experienced by the drug culture of hallucinogenics that provide an artificial mimesis of the electronic information environment. The literate Westerner approaches the study of media in terms of linear motion or sequential transportation of images as detached figures (content), while the right-hemisphere approach is via the ground of environmental media effects instead.

At this point in our discussion, in the context of figure-ground, we come to the core of the matter as far as students of the media are concerned. The basis of all contemporary Western theories of communication—the Shannon-Weaver model—is a characteristic example of left-hemisphere lineal bias. It ignores the surrounding environment as a kind of pipeline model of a hardware container for software content (Fig. 6.1). It stresses the idea of inside and outside and assumes that communication is a literal matching rather than making.

> The *information source* changes this message into the signal which is actually sent over the *communication channel* from the transmitter to the *receiver*. In the case of telephony, the channel is a wire, the signal a varying electrical current on this wire; the transmitter is the set of devices (telephone transmitter, etc.) which change the sound pressure of the voice into the varying electrical current. . . . In oral speech, the information source is the brain, the transmitter is the voice mechanism producing the varying sound pressure (the signal) which is transmitted through the air (the channel). In radio, the channel is simply space, or the aether (if anyone still prefers that antiquated and misleading word), and the signal is the electromagnetic wave which is transmitted.

> The *receiver* is a sort of inverse transmitter, changing the trans-

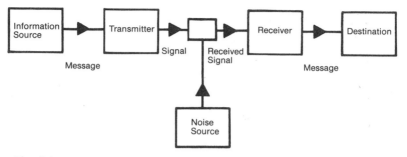

Fig. 6.1.

mitted signal back into a message, and handing this message on
to the *destination*. . . .

In the process of being transmitted, it is unfortunately characteristic that certain things are added to the signal which were not
intended by the information source. These unwanted additions
may be distortions of sound (in telephony, for example) or static
(in radio), or distortions in shape or shading of picture (television), or errors in transmission (telegraphy or facsimile), etc.
All of these changes in the transmitted signal are called *noise*.[7]

Claude Shannon presents his theory of communication in terms
of left-hemisphere verisimilitude as a first object:

The fundamental problem of communication is that of reproducing at one point either exactly or approximately a message selected at another point. Frequently, the messages have *meaning*.[8]

In point of fact, the multiplicity of side effects of any communication system forms an entire environment of interfacings, a kind of
subculture which accompanies the central service of communication. For example, the side effects of the Alaska pipeline were the
subject of a large report by the Berger Commission. The gist of
this report was that the entire native population would be deprived
of its environmental livelihood were the pipeline to be built. In the
same way the side effects of telephone or radio assume a complex
system of electronic technology and supporting services, the adoption of which serves as a new ground that transforms an entire society. Radio bulletins on the hour, for instance, destroyed the *Five
Star Final*. Equally, the system of roads, manufacturers, and ser-

vices that are the side effects of the car have altered the entire face (and odor) of any user society.

We are all trapped in an assumption about the nature of reality and a manner of thinking that has been the hallmark of Western civilization since before the time of Aristotle; the Shannon-Weaver model of communication is simply an extension of that bias. The model and its derivatives follow the linear pattern of efficient cause, the only sequential form of causality in Western philosophy.

Aristotle provides the earliest systematic treatment of causes by drawing together Plato's observations. Aristotelian causality is fourfold, and is applicable both to nature and to artifacts. There are:

> . . . the material cause (the scholastic *causa materialis*), which provided the passive receptacle on which the remaining causes act—and which is anything except the matter of modern science; the formal cause (*causa formalis*), which contributed the essence, idea, or quality of the thing concerned; the motive force or efficient cause (*causa efficiens*), that is, the external compulsion that bodies had to obey; and the final cause (*causa finalis*) was the goal to which everything strove and which everything served.[9]

Originally, the first two causes were regarded as related to how things come into *being* and the last two were associated with the process of *becoming*. All were thought to exist simultaneously. In fact, this doctrine of simultaneous causes lasted until the advent of the Gutenberg era when print (as a ditto device) gave complete ascendancy to visual space. Visual space stresses the detached observer. The detached observer is, ipso facto, placed outside of the frame of experience, and hence the idea of scientific method is born. It is our Renaissance legacy. Have you ever noticed when looking at a Raphael or Caravaggio that the vanishing point produces a certain self-effacement for the viewer? No involvement. (A piazza for everything and everything in its piazza!)

Galileo was ravished by proportional and connected space; he reformulated the definition of efficient cause as the only necessary and sufficient condition for the appearance of something: "that and no other is to be called cause, at the presence of which the effect always follows, and at whose removal the effect disappears." He did not appear to be aware that the right hemisphere, which has today its echo in the instantaneous world of electronic information, involved all of us, all at once. No detachment or frame is possible.

When the concept of Renaissance visual space transformed cosmology and the *logos* alike from resonant ground to rational figure, the understanding of the original idea of formal causality (the structure, essence, or pattern of that which is being realized) was changed from a dynamic to the abstract and the ideal. We have to keep in mind that Aristotle was partially responsible for the confusion because in his own ruminations he had retained and confused both the oral and visual natures of formal cause.[10]

Final cause (that which is the end or purpose of a process), inherent in a thing from the outset, came to be misinterpreted in left-hemisphere terms only as the end point of a whole series of efficient causes. (We must remember that efficient cause refers to learning about something, like a car, by using it.) Formal cause refers to the defining formula or definition of a thing's essence (its form or the "whatness" whereby we know a thing). In other words, at the very least, formal cause and final cause were made subordinate to efficient cause. If it works, it should be allowed to exist.

But prior to the ascendancy of visual space, formal cause was part of a broader spectrum of related considerations; it intersected with *logos* as a figure-ground concern with the entire thing brought into being, structurally inclusive of the whole pattern of side effects on the ground of the users. The first question that could be asked was not whether it was possible to create something, but whether it was desirable in human terms.

In the left hemisphere, formal cause is translated into a kind of Platonic abstract ideal form that is never perfectly realized in any material example. Such is the understanding of Northrop Frye, one of the principal modern exponents of Platonic and Aristotelian ideas as passed through Freud and Jung. He is consistent in his left-hemisphere approach. Referring to the Jungian doctrine of archetypes, Wimsatt and Brooks comment that,

> For Northrop Frye the discovery points to the possibility of turning literary criticism for the first time into a true science. No true science, he argues, can be content to rest in the structural analysis of the object with which it deals. The poet is the only *efficient cause* of the poem, but the poem, having form, has a formal cause

that is to be sought. On examination, Frye finds this formal cause to be the archetype.[11]

Frye is adamant on the point:

> An original painter knows, of course, that when the public demands likeness to an object, it generally wants the exact opposite, likeness to the pictorial conventions it is familiar with. Hence, when he breaks with these conventions, he is often apt to assert that he is nothing but an eye, that he merely paints what he sees as he sees it, and the like. His motive in talking such nonsense is clear enough: he wishes to say that painting is not merely facile decoration, and involves a difficult conquest of some real spatial problems. But this may be freely admitted without agreeing that the formal cause of a picture is outside the picture, an assertion which would destroy the whole art if it were taken seriously.[12]

There is absolutely no provision in Frye's statement for ground of any kind: the archetype is itself a figure minus a ground, floating around devoid of its original context. Otherwise, it would be perfectly natural to observe, along with the rhetoricians and the grammarians, that the formal cause of the poem, painting, or whatever is to be located in the ground. In this sense the ground is the audience (user) and the configuration of sensibilities in the culture at the time the artifact was produced. During World War II, clandestine radio operators on both sides of the fray were frequently more concerned about the effect of the message being sent than its actual content. The message could be false or true depending on the intent of the sender. Agents in the field were known to send back consistently false messages according to prior instruction from headquarters. For a while the entire German spy network in Britain was run by Great Britain and her allies constructing a message environment, through double agents, which appeared to be true. It was a climate of disinformation geared entirely to the users, the Germans.

The four causes as a mode of exegesis of nature had been regarded as parallel to the four levels of interpretation of scripture by medieval grammarians (e.g., Saint Bonaventure).[13] In each case, the "fours" were simultaneous, and both systems were rendered obsolescent by the Renaissance push into visual space and left-hemisphere dominance. Figure-ground resonance and the interplay of levels and causes were eliminated. This had the further advantage, from the standpoint of those who wanted to break from the past, the *moderni,* of cutting all bondage and allegiance to the traditions of the monkish scribe. Bunge summarizes the practical left-hemisphere advantages of dumping manifold causality:

> Some of the grounds for the Renaissance reduction of causes to the *causa efficiens* were the following: (a) it was, of all the four, the sole clearly conceived one; (b) hence it was mathematically expressible; (c) it could be assigned an empirical correlate, namely; an event (usually a motion) producing another event (usually another motion) in accordance with fixed rules; the remaining causes on the other hand were not definable in empirical terms, hence they were not empirically testable; (d) as a consequence, the efficient cause was controllable; moreover, its control was regarded as leading to the harnessing of nature, which was the sole aim of the instrumental (pragmatic) conception of science. . . .[14]

Hence, all Western scientific models of communication are—like the Shannon-Weaver model—linear, logical, and sequential in accordance with the pattern of efficient causality.[15] These are all in the figure-minus-ground mode of the left hemisphere; and, in contrast, do not relate to the instant effects of simultaneity and discontinuity and resonance that typifies one's experience in an electronic culture. The nature of sequential time was such, for example, in the eighteenth century that it was possible to "wait and see." George Washington once remarked, "We haven't heard from Benjamin Franklin in Paris this year. We should write him a letter."

For use in the electronic age, a right-hemisphere model of communication is necessary, both because our culture has nearly completed the process of shifting its cognitive modes from the left to the right hemisphere, and because the electronic media themselves are right-hemisphere in their patterns and operation. The problem is to discover such a model that yet is congenial to our culture and its residua of left-hemisphere orientation. Such a model would have to take into account the apposition of both figure and ground (left and right hemispheres working together and independently when necessary) instead of an abstract sequence or movement isolated from ground.

THE GLOBAL EFFECTS
OF VIDEO-RELATED
TECHNOLOGIES

Global Robotism:
The Satisfactions

Robotism, or right-hemisphere thinking, is a capacity to be a conscious presence in many places at once. It is a right-hemisphere mode—the dominant brain mode of the extended mechanical abilities of our bodies, keyed to one time and one place. Communication media of the future will accentuate the extensions of our nervous systems, which can be disembodied and made totally collective. New population patterns will fuel the shift from smokestack industries to a marketing-information economy, primarily in the U.S. and Europe. Video-related technologies are the critical instruments of such change. The ultimate interactive nature of some video-related technologies will produce the dominant right-hemisphere social patterns of the next century. For example, the new telecommunication multi-carrier corporation, dedicated solely to moving all kinds of data at the speed of light, will continually generate tailor-made products and services for individual consumers who have pre-signaled their preferences through an ongoing data base. Users will simultaneously become producers and consumers.

Nineteenth-century America concentrated on the uniform ethos of a smokestack economy: to be specialist, isolated, and self-directed in its world aims. Extractive industries and agriculture held dominion. A left-hemisphere sense of significant order held sway.

The U.S. population was relatively small and determined to spread itself as far west as possible. Like James Fenimore Cooper's Leatherstocking, Americans were always moving over the hill, through the forest, to the next clearing.

Twentieth-century America, from now until about 2020, will not be engaged single-mindedly in raising crops or throwing up steel mills as much as nurturing people, in an inner-directed way, largely as a result of legal and illegal immigration. Military adventures in Japan, China, Southeast Asia, and Central America have brought about and will continue to be the source of continuous migrations to the American mainland, which will splinter the white, Anglo-Saxon cast of U.S. government, education, and business structures and create a salad-like mélange of ethnic minorities without any single one being predominant.

The recipients of this racial trek will be the supercities of the West Coast and Atlantic South, cities which have doubled or tripled in size as the United States has passed through its century-old movement from country to city and air-conditioning has made year-round work possible. (For example, the Los Angeles–Long Beach area should grow from 3.2 million to 10.1 million in 2033. Dallas–Fort Worth should go from 3.6 million to 7.7 million.*)[1]

Many extractive, agricultural, and low-level manufacturing industries—largely due to high labor costs—will be lost to Third World countries, transforming the United States and some parts of Canada into hard-scrabble competitors in the making of "high-ticket" consumer goods, like consumer robotics and electric commuter cars. While a segment of the U.S. population will be educated and mentally attuned enough to become participants in high technology, most native-born Americans will be unprepared for the new consumer economy which will emerge, offering service-related jobs not always suited to their intelligence or training. Ethnic diversity will help to ignite a full-blown economy based on information exchange.

The Chinese, Japanese, Koreans, Arabs, Lebanese, Mexicans, Central Americans, and Indians who are washing up on U.S. shores by the tens of thousands, legally and illegally, will be well served by the new media technologies. Hundred channel cable

* The metropolitan parameters of Los Angeles, including Los Angeles and Orange counties, had already expanded to 13.1 million by July 1986.

systems will be divided up by culture and language. (Already a hundred and seven languages are being spoken in Southern California.) Videocassettes and videodiscs will spawn new markets for ethnic music, cinema, and stage productions. Regional banks will employ electronic means to create new lending and accounting methods geared to minority traditions of handling money. Neighborhood schools, as in the last century, will be tailormade linguistically. Whether rich or poor, the new ethnics, largely as a backspin against too rapid assimilation, will develop complex and self-integrated barrios.

Although most third- and fourth-generation Americans will be numbed by the coming changes, government and business leaders, with recent foreign backgrounds, will be quick to recognize one inescapable fact about U.S. cities: whereas, in the past, they were primarily transfer and warehousing points for railroad, air, and sea trade, by 1994 many principal cities will be a gestaltic political conglomeration of whites, blacks, Asians, and Hispanics fighting with each other for what is left of the economic pie in a nation of a declining birthrate of native-born Americans and an aging white population. In many older cities, like Buffalo and Detroit, the tax base will have foundered due to a loss of trade functions and heavy industry, prompting a furious competition for federal support.

In those cities, the age of acoustic space in politics will surface with a vengeance. Centers everywhere and margins nowhere in a new tribalism. We may very well see ethnic barrios organizing themselves into self-sufficient, electronically coordinated enclosures, where old-style, ward-heel politics will flourish at the speed of light. The *ma* style of each barrio politician will be his ability to reduce conflicts within his own group and mitigate abrasions with other minorities, maintaining a carefully cultivated separatist image to the rest of the community.

After a generation or two, physical proximity should give way to electronic proximity as the new ethnics intermarry and travel to more remote parts of the country. They will want to keep their parental roots as well as go with the flow of assimilation. Hence, one may expect the construction of special electronic data services to fulfill that need.

The new immigration will help fuel an economic and political surge in the United States and Canada, during the next fifty years, that will have as its ground, or sub-environment, the so-called information age. Computers and sophisticated telecommunication systems should combine to produce work for 80 percent of the population, making the transition complete from an economy based on heavy industry to a marketing, service-oriented economy, having the needs of the singular consumer at its center. But as we have indicated in the first half of this book, the essential change in the U.S. will not occur so much in the proliferation and diversity of tailored artifacts as in the minds of the men and women who will produce them.

The United States by 2020 will achieve a distinct psychological shift from a dependence on visual, uniform, homogeneous thinking, of a left-hemisphere variety, to a multi-faceted configurational mentality which we have attempted to define as audile-tactile, right-hemisphere thinking. In other words, instead of being captured by point-to-point linear attitudes, so helpful to the mathematician and accountant, most Americans will be able to tolerate many different thought systems at once, some based on antagonistic ethnic heritages. Social patterns will have more weight than alphanumeric measurements. But, nevertheless, we cannot look for a balance between the hemispheres right away. Having few ethnical, social, or conscious restraints, America is destined to plunge headlong into right-hemisphere values and attitudes, perhaps abandoning for a time the virtues of precise naming and quantitative ordering, much as some bright teenagers desert their studies for the uncertain joys of covert data-base searches and bootleg videogames. One could make a case that twenty-five years of television viewing has already set the stage for this psychic shift when one considers that the average North American family spends seven and a half hours a day in front of the cathode ray tube, to the neglect of more stimulating activities.

The real meaning of the legend of Narcissus is that he did not fall in love with an image of himself but rather the face of a seeming stranger. Zeus made him gaze into the watery pool which gave back a reflection of someone *like* him but different enough to be fascinating. Not replica but re-presentation. This is precisely what happens when we project our bodily and psychological functions onto the world outside. We "amputate" them because we cannot gaze too long at a balefully realistic playback of ourselves. To some extent, the function of art is to provide some livable distance.

All media are a reconstruction, a model of some biologic capability speeded up beyond the human ability to perform: the wheel is an extension of the foot, the book is an extension of the eye, clothing an extension of the skin, and electronic circuitry is an extension of the central nervous system. Each medium is brought to the pinnacle of vortical strength, with the power to mesmerize us. When media act together they can so change our consciousness as to create whole new universes of psychic meaning.

That glowing phosphorescent eye sitting altar-like at the end of the living room is no exception. Herbert Krugman, in experiments conducted for General Electric, seems to have been the first man to discover the relationship between television and the alpha state. Picture yourself sitting down for a night's viewing. You have had a day's worth of analytical problems, whether you have been fixing a car or doing actuarial tables. You switch on the set. Almost immediately your left brain slides into a nondominant, neutral state, lulled by the dots flashing sequentially across the screen at one-thirtieth of a second.[2] But the right brain remains alert stimulated by bright, sensuous images, music, and random movement. The right hemisphere may be the seat of emotion; and, if not that, then handily connected to those limbic regions which give forth a tympany of sub-primate responses from below the neo-cortex. Freed from the restraints of the watchdog left, your mind is in a condition to respond to virtually any suggestion, especially of a sensuous or symbolic nature, and you are fair game for the nonrational sell.

The home may very well become more efficient and automated as cable TV, videocassettes, videodiscs, and quadraphonic sound are added to new home construction. For those who need escape, high-density screens will amplify and accentuate the alpha state. For those seeking information, TV linked to the computer might eventually surpass the resources of the Library of Congress. The speed of print data through satellite hookups, such as Associated Press Newscable, could deliver to individual users an overwhelming range of information fashioned, perhaps, to one's professional needs. The possibility of constant live information would prompt a continual update of background data on key news events. Audiences oriented to a videogame mentality, neglectful of books and newspapers, might over a period of time welcome a capsule style of reporting, which when pushed to its farthest limits reverts to the style of the ideograph.

In the one-way "distributive" mode, television—if it remains in the hands of the white, Anglo-Saxon establishment—could become a buttress blunting the disruptive effects of ethnic diversity. In the two-way, "interactive" mode, individual data-base users could utilize the medium to resist the propagandizing character of national network programming, since it emanates from only two or three sources. In either case, the records kept by cable system owners will no doubt be used to construct profiles of purchasers' habits and opinions; which, in turn, will be sold to merchandisers who will solicit buyers for products.

By the power of this combination of video-related technologies alone, the U.S. economy would be finally shifted from a manufacturing to a marketing society. Most of the telecommunications investment in America is now going into land-bound cable with the object of achieving an 80 percent coverage of U.S. homes by 1990.[3] Cable, whose chief technical ingredient is coaxial lines, is currently devoted to refining and repeating the signals originated by the national diffusion networks (ABC, CBS, and NBC), but its most important function, yet to be fully tapped, lies in its "narrow-cast" two-way quality: that is, its capacity to send signals from the cablehead and, at the same time, receive signals from individual homes and businesses.

When fully realized these interactive abilities can, at least in the beginning, be used to accomplish such things as routine housekeeping chores, security, and teleshopping. At the very least it means more personal freedom for the householder and the chance to work at home. This is another way of saying that the home could become a center point once again in American society, as it was on the frontier. Serial marriage and divorce will produce all manner of extra relatives and half siblings, making the family seat a shelter for the extended clan and a powerful pastiche of shared psychological and economic interests. Add to this the fact that the U.S. population is becoming steadily older on average (over 20 percent), and a new nub-point of conservatism could clearly emerge in future.*

What will happen in the home will become implicit in the workings of the information-service economy. The consumer as producer will take the initiative away from the conglomerate.[4] In the nineteenth century, there appeared a vertical organization of society geared to raw materials, manufacturing, and distribution, laid out geographically with the railroad acting as a sort of connective tissue. Towns and cities—often called company towns—grew up and drew their life's blood from one of those hierarchical vertical activities. Buffalo, New York, was a steel town; Roseburg, Oregon, devoted itself to lumber.

In the information age, however, we shall see whole regions devoted to a balancing combination of industries in the same sense that "Silicon Valley," south of San Francisco, is keyed to all the products of photonics and microelectronics and the Orlando area revolves around the transportation, travel, and tourism complex of Disney World. Industry in the twenty-first century will be horizontally affiliated. The computer, working at the speed of light through a myriad of communication devices, will produce tailor-made products and services for potential buyers who have already presignaled their preference through the database, whether it be a perfectly adjusted insurance/investment program or a dream vacation.

One can easily visualize the development of affiliated corporations dedicated to the national management of finance and insurance, construction, mining, hi-tech manufacturing, agriculture, and utilities. At first, computer hardware will intensify centralism. But as soon as the software is developed which emphasizes the needs of pre-registered clients, the result will be decentralizing. If the personal and operating data of each household could be contained in a data base, the affiliated corporation (with the permission of its clients) could design, build, and control all home utilities, housekeeping records, general payments, and tax filings for millions of people, regardless of location. Freed to pursue other interests and diversions, the client might very well consider such services well worth paying for.

---

* More precisely, as of 1988, the number of U.S. persons over 55 was 51.8 million (21.3 percent of 245 million), and the number over 65 was 29.8 million (12.2 percent).

Most of the troubled inner cities of the Northeast and Midwest will probably not be able to attract the affiliated service corporations. High taxes, crumbling utilities, cramped living space, and intense crime areas all may be bypassed in favor of suburban and rural areas, in much the same way the Prudential Insurance Company chose to place its axial computer center in Roseland, New Jersey, and Harcourt Brace Jovanovich sited its new corporate headquarters between Tampa, Orlando, and Daytona Beach.

The affiliated corporation (AC) should be a direct product of the current state of signal transmission. Copper is a costly mineral which has, on occasion, been difficult to obtain since World War II. Fiber-optic glass, on the other hand, is made from one of the commonest elements in the universe, silicon; and one pound of fiber-optic strands yields up to eighty times more data (analog and digital) than a pound of coaxial cable.[5] It is uncommonly light and flexible. But, more important, fiber optics can make possible the low-cost and efficient linking of all of the terminal hookups necessary for an up-to-the-second orchestration of sound and visual information (through infrared light)—which cannot currently be done with copper wire, microwave, and coaxial cable. Fiber optics constitute the electronic backbone of the DACS computer center at Disney World, where "all aspects of show performances 'onstage' throughout Walt Disney World (27,000 acres) are monitored from the opening and closing of theater doors to the singing of bears, birds, and the speeches of pirates and presidents"—and, one might also add, the direction of all lighting, heat, air-conditioning, and garbage disposal.

The AC, in other words, whether public or private, is an electronic "omnium-gatherum" which unlike the products of the machine age is keyed to human rhythms—a reciprocating dialogue with the environment emanating from the central nervous system; an artifact unique to our century. The satellite cum data base gives to such organizations as the Associated Press, Walt E. Disney Enterprises, and Citicorp, a New York bank holding company, the power to adjust all their operations from pole to pole in one-fifth of a second. The major video-related technologies (fiber optics, computers, microwave, and satellite) obliterate distances, to be sure, but on an interactive basis. This computer/data-base power of simultaneity will cause the literal implosion of some businesses and public services, which is the essence of robotism.

Writing was a device to govern by paper over long distances. Calendars were devised, like clocks, to trace the passage of written and printed messages from one region to another and to centralize the organization chart. Instant control, however, eliminates the middle man. There is, for example, no technical reason why the 40,000-odd financial institutions in North America devoted to banking, securities, and insurance could not be merged into a single institution through electronic means. Horizontally arranged, multi-service ACs could become regional and global, like Diner's Club or American Express, merely through international charter. The ability to organize on a worldwide level at low cost would give some affiliated corporations more power than any single international business or modern state. For instance, Citicorp today (with its 90,000 people in 3000 offices in 90 countries) through the manipulation of whole currencies could, if it chose, bring about the fall of governments.

The mature affiliated corporation will have no irreplaceable administrative locus; it will be structured acoustically with many centers. The Associated Press has ten major computer banks placed in a hub pattern around the United States. If one were to fail, the other banks can be made to share automatically the message load. Accordingly, the AC of the future will assume a spherical character, like the Bell System telephone network. AC managers at teleports in world locations would sit at multiplexing "organs" bringing time, space, and satellite frequencies together as a resonating whole—time and space undivided, in the sense that one cannot divide a musical note.

Global Robotism:
The Dissatisfactions

Robotism is also decentralizing. The invention of the alphabet and writing tended to complement the ancient propensity to concentrate, in a sedentary way, power and resources. The scribe had a strategic left-hemisphere position in centralized bureaucracies, well into the twentieth century.

In an electrically configured society all the critical information necessary to manufacture and distribution, from automobiles to computers, would be available to everyone at the same time. Espionage becomes an art form. Culture becomes organized like an electric circuit: each point in the net is as central as the next.

Electronic man loses touch with the concept of a ruling center as well as the restraints of social rules based on interconnection. Hierarchies constantly dissolve and reform. The computer, the satellite, the data base, and the nascent multi-carrier telecommunications corporation will break apart what remains of the old print-oriented ethos by diminishing the number of people in the workplace, destroying what is left of personal privacy, and politically destabilizing entire nations through the wholesale transfer of uncensored information across national borders via countless microwave units and interactive satellites. The twenty-first century will be the age of *Aquarium,* by common consent. Left-hemisphere thinking will atrophy, submerged in acoustic space.

Ecology shifts the "White Man's Burden" onto the shoulders of the "Man-in-the-Street." The meaning of the atomic bomb is that we can no longer fight territorial wars as a sort of solo game-playing, so beloved of the left-hemisphere. The age of information remakes the world in our image. The media extensions of man are the hominization of the planet; it is the second phase of the original creation.

Territorial fights spring from a sense of isolation by interval (the illusion of mechanization), which is a transformation of some aspect of nature, or of our own bodies, into amplified and reconstituted forms. We extend parts of ourselves out into the environment to do some intensely elevated function (i.e., wheel (feet), hammer (fist), knife (teeth-nail), drum (ear), writing (eye)) and then find ways to fight about it. The early ape with a club was a specialist. The first humanoid uttering his first intelligible grunt, or "word," outered himself and set up a dynamic relationship with himself, other creatures, and the world outside his skin. Speech entails competition. It is also a tool to reconstitute nature into working synthetic models, to translate one form into another. Conflict occurs, not because of human inefficiency, but technology moving at incompatible speeds.

The pre-neolithic art of making stone tools moved man out of the process of evolution and into a world of his own making.[1] The hunter became the neolithic planter. Being in one place gave man the opportunity to count the ways into which he could divide himself. Early on, as a pre-literate he outered his whole body into a ship or a hut or rollers (the Incas had no wheels). As a presumptuous literate, having regimented the character of thought (mental dance, nonverbal ESP), he specialized by outering only portions of himself, imitating the sequential mental climate of the phonetic alphabet. He unrolled his uniform sense of wholeness and sectioned himself like a salami. Power came through amplified physical repetition.

Whether maneuvering his coracle, bow and arrow, battle tower, or steam engine, the translation of these media (and his robotic relation to his own inventions) was only partial, the extension of one sense at a time. Nevertheless any medium, by dilating a particular sense to fill the whole field, creates the necessary conditions for hypnosis in that area. The medium becomes an unknowable force to the user. This explains why all societies are initially numbed by the adoption of any new technology. At no time in man's history has any culture been aware of the effects of its outered media upon its overall associations, not even retrospectively. The Tartar stirrup created the medieval knight as tank; a fact that astonished the successors to the Huns.[2]

As man succeeds in translating his central nervous system into electronic circuitry, he stands on the threshold of outering his consciousness into the computer. Consciousness, as we have discussed in a previous chapter, may be thought of as a projection to the outside of an inner synesthesia, corresponding generally with that ancient definition of common sense.[3] Common sense is that peculiar human power of translating one kind of experience of one sense into all other senses and presenting that result as a unified image of the mind. Erasmus and More said that a unified ratio among the senses was a mark of *ratio*nality.

The computer moving information at a speed somewhat below the barrier of light might end thousands of years of man fragmenting himself. Up to now, the extensions of man have been warring with each other: spear against gun, stagecoach against railroad engine, television against radio, at incompatible speeds. The horizontally organized, multi-service corporation, or something like it, in its use of information as wealth by electronically predicting consumer needs before the first wheel is turned or button pushed in factory or retail outlet, may be returning us to a state of integral awareness.

We are entering the age of implosion after 3000 years of explosion. The electric field of simultaneity gets everybody involved with everyone else. All individuals, their desires and satisfactions, are co-present in the age of communication. But computer banks dissolve the human image. When most data banks come together into a reciprocating whole, our entire Western culture will turn turtle. Visualize an amphibian with its shell inside and its organs outside. Electronic man wears his brain outside his skull and his nervous system on top of his skin. Such a creature is ill-tempered, eschewing overt violence. He is like an exposed spider squatting in a thrumming web, resonating with all other webs. But he is not flesh and blood; he is an item in a data bank, ephemeral, easily forgotten, and resentful of that fact.

Earth in the next century will have its collective consciousness lifted off the planet's surface into a dense electronic symphony where all nations—if they still exist as separate entities—may live in a clutch of spontaneous synesthesia, painfully aware of the triumphs and wounds of one other. "After such knowledge, what forgiveness." Since the electronic age is total and inclusive, atomic warfare in the "global village" cannot be limited.

As new technological man races toward this totality and inclusiveness, he will no longer, as in earlier times, have an experience of nature, as "nature-in-the-wild." He will have lost touch, and by now we should realize that touch is not simply skin pressure but a grasp of all senses at once, a kind of tactility. When we lose nature as a direct experience we lose a balance wheel, the touchstone of natural law. With or without drugs, the mind tends to float free into the dangerous zone of abstractions.

Arnold Toynbee wrote that incompatible societies will always fall into a confrontational situation with each other, that a complex civilization, for example, growing rapidly beside a less-developed, tribally oriented group will rain down a blizzard of psychic suggestions as a counter-irritant which will inevitably result in an explosive reaction. This observation, played in reverse, tells us that the inner-directed person, especially one inflated with an almost Emersonian view of individualism, will be emasculated by the effects of acoustic space because he is not trained to perceive it.

In this century the Third World has increasingly been manipulating the West. "Weaker" societies invade and conquer "stronger" societies not by arms but through infiltration in much the same way the people of the Southern Hemisphere and the countries of the Pacific Rim have been slipping into the United States because the white, Anglo-Saxon majority has been unable to "see" them. Right-hemisphere-oriented people, like the African Blacks, are invisible to those who cannot think in qualitative terms. When the Banana Republics began to destabilize over land reform in the twenties and thirties, the U.S. reaction was predictably lawyer-like and aggressive, a call to the military to make those "greasers" behave.[4]

Like education and industrialism, the military of the West is the product of the homogenizing effects of the phonetic alphabet, King Cadmus's Dragon's Teeth. Occupation to the U.S. Marines was a container to be filled, not a process to be monitored. The people of Central America absorbed the gringo thrust and blunted it with a lotus-like effectiveness. The multitude has no use for time laid out in intervals, keyed to a demand for results. Only specialists think that way.

The person who gives over his life to electronic services, whether he is merely a participant in a cable system or an information manager, will lose the security that proceeds from specialism. Specialism developed in the Western world as a reaction to the new social order devised by Solon for his fellow Greeks. Henceforth, proclaimed the lawmaker, the Athenians will make goods only for export, leaving the agricultural bias of the Attic Plains to itself. Soon the Greeks added foreign slaves and profits soared. They began to entertain the idea of a job as a repetitious assembly-line method of making goods, which is undoubtedly the source for the Greek word *tekné,* art, or made-by-hand.

The idea of the role was gradually lost sight of—that is, the multiple holding of partial jobs signifying one's authority over a household. The specialist can always be seen to have one salient characteristic: he is quite willing to trade his freedom of action for the security and the stability of a closed system. Odysseus undoubtedly felt the sting of this commitment after returning home to Penelope, climaxing ten years of creative wandering. Toynbee explains that in a culture of active warriors, the lame and the crippled (and the old) become specialists, like Haephaestus, the smith and armorer.

The Russians have never really moved beyond specialism. Russian austerity is founded on the fear of the new media and their capability of transforming social existence. The Russian revolution reached the stage of book culture. The filmmaker Sergei Eisenstein was tolerated but his images generally suspected. Lev Kuleshov began his career as a film editor by making documentaries that seemed like biologic text records. The Russian party line stands pat on the status quo ante 1850 that produced Engels and Marx. Karl Marx never studied or understood causality. He paid no attention to the railway or steamboat.

Present-day Russians strive to live within the nineteenth century of consumer values, allied to the idea that the state ownership of the means of production really has a crucial effect on society. Accepting the possibility that, through electromagnetic means the mass consumer without owning anything could become the czar of production, seems quite beyond the current crop of party theorists. And beyond the urbs of Moscow and Leningrad, with their tenuous grasp of Western property ideas, lies the rest of the nation: tribal, corporate, and nonvisual.

What may emerge as the most important insight of the twenty-first century is that man was not designed to live at the speed of light. Without the countervailing balance of natural and physical laws, the new video-related media will make man implode upon himself. As he sits in the informational control room, whether at home or at work, receiving data at enormous speeds—imagistic, sound, or tactile—from all areas of the world, the results could be dangerously inflating and schizophrenic. His body will remain in one place but his mind will float out into the electronic void, being everywhere at once in the data bank.

Discarnate man is as weightless as an astronaut but can move much faster. He loses his sense of private identity because electronic perceptions are not related to place. Caught up in the hybrid energy released by video technologies, he will be presented with a chimerical "reality" that involves all his senses at a distended pitch, a condition as addictive as any known drug. The mind, as figure, sinks back into ground and drifts somewhere between dream and fantasy. Dreams have some connection to the real world because they have a frame of actual time and place (usually in real time); fantasy has no such commitment.

At that point, technology is out of control. The Greeks very early lost control of technology when they substituted the idea of the private citizen and written legal codes for the peer wisdom of traditional communities. During what we identify as the Golden Age of Greek literature, Herodotus remarked that his people were "overwhelmed by more troubles than in the twenty preceding generations. . . ." In the Western world we are heading for an inrush of social aims and structures. The group mind will predominate and make us so sensitive to other people's needs and wants that whole regions will be exhausted by the demands of adjustment.

But more deadly than minute and constant calls for change, especially when those affected are unaware of its cause, is the attitude of mind which has persuaded Western man to take on the duties of a god. *Sputnik* in encircling the planet made it an object of art. That small aluminum ball called forth a view of the earth as something to be programmed. Like the pilot of the space shuttle, man is now captain of spaceship earth, engendering a concept of ecology—of earth, air, fire, and water as an integrated whole. There are no more passengers, only crew. Such a grasp of totality suggests the possibility of control not only of the planet but of change itself. Constant change, for its own sake, threatens everybody. (One of the interesting things about continuously mutating technology is that it is one of the prime sources of inflation.)

In a state of social implosion, induced by information moving at the speed of light, those who are part of information monopolies, like the foreign-exchange analyst or the book editor, will not see change as threatening. But when ordinary people do not know who they are, they get anxious and violent. Many men went to the frontier in the last century to prove themselves. In the border town of the American West, everybody was a nobody until he wrested an identity through taking a risk and pure grit. The frontier was a hardware society which allowed men and women to define themselves by transforming the land.

The electronic society does not do so; it does not have solid goals, objectives, or private identity. In it, man does not so much transform the land as he metamorphosizes himself into abstract information for the convenience of others. Without restraint, he can become boundless, directionless, falling easily into the dark of the mind and the world of primordial intuition. Loss of individualism invites once again the comfort of tribal loyalties.

As the well-developed industrial nations of the Northern Hemisphere keep their population growth down through birth control and move toward a socially implosive situation, the nations of the Third World, particularly of Africa and East Asia, are increasing their populations at an average yearly rate of 2 and 3 percent. In a five-year period (ending in 1980), India added nearly 50 million people. Kenya may double its population in the next twenty years, from approximately 23 million to 45 million.* By the year 2020, when world population is reassessed, the Third World will hold

---

* Kenya may double its current population by the end of the century according to United Nations sources; by 2025, this nation, which includes a major part of East Africa, may exceed 80 million persons.

over 80 percent of the planet's people and will have run out of land to feed itself.[5]

Mexico will press against the southern border of the United States, weighing in at 100 million souls.[6] Consider what Spanish-speaking U.S. citizens might do politically and economically to keep migration flowing across the Texas-California frontiers, especially if Hispanics strike up informal coalitions with other minorities to redefine the power of federalism. Super regionalisms and separatisms, like the *Parti Quebeçois* in Canada, argue the possibilities of superstates (and the consequent demise of small countries) in order to cope with world population surges and the prospect of mass starvation.

The electronic planetary surround will dramatize daily the plight of the homeless and starving and the objective, quantitative position of literacy will be everywhere under attack. In the last two decades, when the United Nations Educational, Scientific, and Cultural Organization provided free radios throughout Africa and the Middle East, it shook the foundations of reading and writing in these areas. For over 300 years white missionaries had toiled to detribalize whole nations with the Western alphabet. In little more than a generation, radio (and later TV) has brought the audile basis of the tribe once again into high relief.

In Iran and Libya, Western left-hemisphere values have been toppled. The Mullahs have re-established their rule through the power of the crowd, diffusion broadcasting, and audiocassettes. Temple rule is based on time and the habit of engraving in stone. Military bureaucracy depends on paper and courier systems. Video-related technologies compress the sequent into the simultaneous and emphasize the pre-literate group will, re-establishing the tribal chieftain. Computer programmers will become the new Pythagoreans, espousing pattern as the golden mean.

Christopher Lasch demonstrated in *The Culture of Narcissism* (American life in an age of diminishing expectations) that he did not understand the basic mechanism of figure and ground.[7] When the ground moves too fast, a condition endemic to the electronic society, only figure is left. The left-brain oriented individual substitutes the act of going inside himself for identity. He uses his own figure as his ground. Hence the professional actor's vortical and often self-destructive egotism. At the speed of light you become a narcissist because only the figure of self remains—which explains, as Tom Wolfe has pointed out, why some jet-setters are so involved with themselves. They literally belong to no one community; therefore their community is inside their well-cared-for skins.

Narcissism, as a side effect of acoustic space, is, beside AIDS, the fastest developing social disease of the peoples of the West. Yet, at the same time and certainly by the turn of the century, the Third World will implode upon itself for different reasons: too many people and too little food. When one is concerned about food and shelter for today and tomorrow, it is very difficult to be preoccupied with independent goals or future social choices. The "man-on-horseback" beckons towards Armageddon. The tetrad of the cancer cell reveals, in small, the immediate hereafter of the world: cancer enhances cell reproduction, obsolesces the equilibrium or homeostasis of normal cell production, retrieves primitive cell evolution, and transforms itself into self-consumption. Starvation promotes self-consumption.

In the year 2020, nearly 8 billion people (as against today's 5.5 billion) will crowd the planet;* all but 15 to 20 percent will live in today's undeveloped nations. If the First and Second Worlds (the United States, Europe, Russia, and Japan) wish to avoid a disastrous fight to the finish between the haves and the have-nots, they had better be prepared to provide food and psychic leadership to the entire planet, without regard to national priorities. The new technologic man, hypnotized by his own electronic navel, must become his brother's keeper, in spite of himself.

The role of the shepherd, a continuing archetype in biblical literature, invariably entails a spiritual quest. The wolf prowling ravenously about the flock is God's agent for self-examination. In the near future, our spiritual quest may lead us not, as some have thought, into outer space but into the depths of the sea for additional sources of food and different techniques of survival. But such an exploration of inner space could, at first, take the form of a radical overhaul of educational methods.

Jack Fincher in *Human Intelligence* notes that our school establishment is strongly slanted towards left-hemisphere standards and skills. Under electronic conditions the right hemisphere gets sa-

---

* The United Nations estimate of world popualtion in April 1987: 5.5 billion; in approximately the year 2020 that figure should rise to 7.8 billion.

lience and preference for the first time since the advent of the alphabet 2500 years ago. The electric ground, like the multi-dimensional nature of the ocean, makes an environment favorable to the right hemisphere. From the viewpoint of some teaching psychologists, the male preference for dominance would be displaced by the female quality of nurture.

From babyhood, education might be geared to the biologic developmental stages of the mind. Early testing could determine whether a person learns best in a visual or acoustic mode. At their own speed, boys and girls could be taught such subjects as English, physics, math, and chemistry in either left- or right-hemisphere terms. The opposite case prevailed when mechanical and industrial man created environments of lineal services and qualified markets. Electronic man creates environments that are simultaneous, open-ended, and acoustic.

Paradoxically, electronic man is re-creating the conditions of the Orient and the Third World as the norm for our new world. Instant readjustment to surrounding, or robotism, cannot be avoided. The new passion for Zen and the *Tao of Physics* and ESP has an electronic base which is irresistible, because unconscious. Our left-hemisphere educational establishment, embosomed in the Euclidean architecture of Walter Gropius and Mies Van de Rohe, is dedicated to achieving quantitative goals. The new right-hemisphere society (coming to the West) prefers artistic role-playing and an indulgent enjoyment of the quality of life rather than quantity. The new education will have no goals whatsoever.

Governments need to know that electronic services, especially television, eliminate or dissolve representative government. TV ends representation at a distance and involves one in the immediate confrontation of an image. The successful image will be charismatic, meaning that it represents a great many admirable types. (President Carter was Huck Finn in the White House.) The other side of the image is that it tends to become fantasy (discontinuous "flash") under the discarnate conditions of TV-viewing, that is, the audience wishing to be there floating in the electronic void rather than being fixed at home.[8] For the new popular image, of which Reagan is an exemplar, there can be no relevance in parties and policies but only a war of icons or images. Coupled with the cable-based two-way mircrocomputer, the electronic referendum becomes a species of home rule. No congressman or provincial representative will be able to function without it.

In a word, the difference between the current TV generation and the pre-TV population as precursor of the new age is basically the difference between fantasy and dreams. The fantasy TV generation is not seeking deferred or delayed rewards. The dream generation, on the other hand, which included the radio and movie segments of that population, had dreams and goals based on the star system and the ideals that went with it. Women could model themselves on Greer Garson, and girls, like Judy Garland, could sing to Mr. Gable that "you made me love you." There are no stars, however, in TV, only "television personalities."

Since the basis of natural law is unavailable to the TV generation, its only recourse is to supernatural law as a means of coherence and meaning. The Beatles seek the gurus, and their groupies drift into Hare Krishna. For these reasons, with the fast developing physical plight of Third World as a ground, we appear to be on the threshold of a new religious age—the age of *Aquarium*. But before this all-involving age reaches its apotheosis, four major video-related technologies (computer, satellite, data base, and the horizontally organized, multi-carrier corporation) will break apart what remains of the old phonetically literate society by causing massive unemployment in the industrial nations, a destruction of all privacy, and a planetary disequilibrium keyed to continent-wide propaganda skirmishes conducted through the new-found utility of interactive satellites.

A satellite placed in orbit at 22,300 miles above the earth can situate a nonjamable "footprint" over thousands of square miles, with a clarity long-distance radio could never achieve. When one billion people can be privy to a pope's investiture through world television, one can be certain the prayer mat is about to replace the Cadillac with a new psychic awareness.

A tetrad, as we have said before, is simply an intuitive tool based upon principles very similar to Heraclitean dynamics involving the reconciliation of opposites. The tetrad demonstrates that within each of man's inventions (extensions of himself) left- and right-hemisphere modes of thought struggle for dominance and, in the speed of the electronic age, reveal themselves instantly. A tetradic examination of the four video-related mediums cited above shows their nonexpansive, implosive character leading to total involvement as compared with the expansive, splintering character of artifacts from the mechanical or industrial age.

The computer is the first component of that hybrid of video-related technologies which will move us toward a world consciousness.[9] It steps up the velocity of logical sequential calculations to the speed of light, reducing numbers to body count by touch. When pushed to its limits, the product of the computer reverses into simultaneous pattern recognition (acoustic space), eroding or bypassing mechanical processes in all sequential operations. It brings back the Pythagorean occult embodied in the idea that "numbers are all"; and at the same time it dissolves hierarchy in favor of decentralization. Any business corporation requiring the use of computers for communication and record-keeping will have no other alternative but to decentralize. When applied to new forms of electronic-messaging, such as teletext and videotext, it quickly converts sequential alphanumeric texts into multi-level signs and aphorisms, encouraging ideographic summation, like hieroglyphics (see Fig. 8.1).

Computers are designed for simple quantitative speedup, that is, to do the repetitive reading, writing, adding, multiplying, and dividing that we all get bored with—a kind of mind-numbing that frequently engenders errors. A computer never gets bored; it thrives on monotony. It doesn't make mistakes. (The people who instruct it make mistakes.) It can get "sick" from time to time and require mechanical rehabilitation. But, in general, the computer fosters accuracy and consistency. These qualities are best demonstrated in record-keeping and data accumulation. And it is exactly in these functions where the effects of the computer on humans can be perceived.

First of all, it is important to understand that the workplace is the chief living area of most Canadians and Americans. We spend more time there than at home. The timetable and the goals of the workplace, whether we are talking about banking, car assembly, office equipment manufacturing, warehousing, or the making of foodstuffs, often overwhelm the priorities of the home. Indeed, the people we work with at times have more emotional impact on us than our own families, because in the workplace, like the frontiersmen of old, we fight our private singular battle for survival. That

Computer

A. Accelerates
calculations to
speed of light

D. Reverses into
simultaneous pattern
recognition

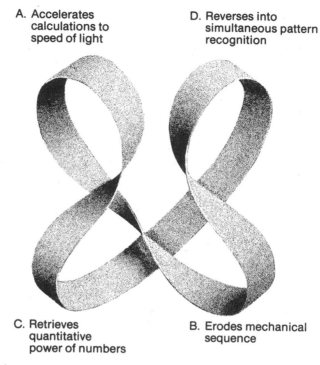

C. Retrieves
quantitative
power of numbers

B. Erodes mechanical
sequence

Fig. 8.1.

battle requires us to have allies, a few trusted friends, who become necessary for information and support. The day-to-day interaction among secretaries, clerks, workers and managers, the social life, if you will, among people in the workplace, is the matrix against which the work of the business is done.

Computers will, in the long run, dramatically alter the social environment of the workplace as we know it. There is no point trying to avoid what is coming, because to a great extent that structure of change is already here. Let us, for example, examine the nature of that change in most "front office" situations, that is, organizational units which collect and analyze information for decision-making purposes. Richard Crump, a management analyst for Northern Bell, developed a paradigm for analyzing most work-place conformations. He says that information workplaces have three kinds of people: the processors, the concentrators, and the interactives.

A processor generally does nothing with data material; he or she simply translates it from one medium to another. For instance, a clerk in the stock exchange could spend all day transporting numbers and symbols from phone to paper or paper to CRT. The essential thing is that the processor does not add or subtract from the total data available.

The concentrator collects, prepares, and partially arranges data for someone else's use. It is usually done for a particular purpose. The meteorologist will collect temperature and barometer readings in order to outline high and low areas on the weather map. Monthly sales figures will be routinely compiled for comparison at a later date. Subject matter varies. The concentrator is usually someone who can work alone without reference to other people, needing, however, occasional direction.

The interactive type is a manager—the social center of the office situation, representing the organization to outsiders and regulating the work force according to specific objectives. Essentially, he or she decides how to manufacture the raw data into the finished product of information. The interactive person is charged always with the responsibility of collecting data, passing it through a series of planned actions to produce a set of results. From the quantitative, as it were, to the qualitative.

Why is all this necessary? Those who run any business, whether it be a university or a manufacturing plant, need information, as opposed to raw data, to control the operation. They need to keep track of what is going on between the internal life of the organization and the outside world—essentially to keep track of the number and kinds of transactions so that they know what they are accomplishing and can further assess what to do. Transactional records have to be assembled in one place (data base) and related to current business. Current business, to be comprehensible, has to be related to the past so that statistical materials can be assembled for accounting, billing, customer advice, and periodic reports, both to management and government. The problem was the same for the high priests in charge of the Babylonian granaries as it is for us today. Only the tools have changed: from abacus to electronic calculator, so to speak.

The average business computer largely mimics the functions of the processor and the concentrator. It collects raw data from the data base, sends it through a series of planned actions via a program, and produces information which, in this instance, can be defined as "news you can use." The program is a list of instructions devised by the interactive person; a "recipe" to produce timely facts needed for review and control.[10] In the future, the orchestration of such facts will become more important than ever before. As facts move at the speed of light, the techniques of capturing them analytically, in time and space, will be the special province of information specialists who will harmonize machines more than people. In effect, the processors and concentrators will be replaced by electronic circuits.

It won't happen right away; the transition will be gradual, yet inexorable. Perhaps a better way to understand this process is to realize that the innards of the data-processing computer, which has essentially not changed in principle in the last thirty-six years, is a simulation of general office behavior. Parenthetically, what really has changed is the speed of computation, from a thousandth of a second to a trillionth of a second. The comparison is broad, but nevertheless clear.

In most office functions, there are people who spend a good part of their working day taking or checking orders; once checked these materials are often passed on to other departments (the processors). There is another group which assembles current transactional data for updating and billing purposes, past and present (the concentrators). All these functions are controlled by a chief clerk usually under an office manager. In every computer, which performs data-processing tasks and has a logic capability, the job of the processor is replaced by input and output elements. The job of the concentrator is taken by memory and data storage. The agency of the chief clerk is taken by the central processing unit. The really operative person is the programmer-manager or systems analyst who devises the original electronic directions to synchronize the computer "office." In short, the entire operation has been miniaturized, speeded up, and placed under the direction of one mind instead of several.[11]

Therein lies the problem. One person, one might imagine, sitting before a terminal, mesmerized by the product of the human mind, believing perhaps that the brain is as "perfect" as the machine. The computer is a massive enlargement of only one level of reasoning—what philosophers are prone to identify as efficient cause (cause and effect). It deals with only "yes" and "no," the essence of the excluded middle, the digital form. It allows no consideration of opposites of equal power. The plan of the computer allows no other form of reasoning and cannot ask questions about the antecedents of its own programming. Hence, the programmer-manager is in constant danger of becoming a self-mesmerized robot. But more than that are the warnings of social isolation. Nathaniel Hawthorne spent a lifetime telling us how perilous it is to have no other spiritual or moral measure but our own.

The very nature of the computer will push logical (mathematical) maturity to the point of breakdown. Most logical sequential calculations can easily be driven to the speed of light. As this process evolves, it will bring back and accentuate an ancient preoccupation with the mystical quality of numbers in a sensuous tactile mode.[12] At this stage of greatest intensity of development, there will be an unanticipated reversal: the simultaneous will emerge from the sequential, the mythic from the historic, acoustic from visual space. The old ground rules of point-to-point logic will break down. And holism will then emerge as a dominant form of thinking, governed by a considerably smaller group of management elite.

In the same way that no telephonic engineer can conceive of an entire coast-to-coast network in anything but a 360-degree dimension, software—in the computer—will in itself become an art form, with individual highly recognizable human signatures. Human signatures, however, can be easily counterfeited. Hence, one of the elements of holistic thinking in the future will be the need to encrypt whole data networks and satellite systems to protect key areas of information, without which corporations and governments could not function.[13] A small elite will become the guardian at the temple gate.

The data base is an electronic library capable of massive storage and instant accessibility. The storage capacity of the first electronic computer, ENIAC, can now be placed on a microchip. The data bank, coupled with the computer and serviced by a variety of electric transmission networks, can be made to record (for simultaneous retrieval) all one's financial, social, educational, and personal transactions from birth. Major data bases are abuilding all over the Western world, but to date none has reached the mature dimensions (with the possible exception of those of the Internal Revenue Service and the Federal Bureau of Investigation) of the banking industry's Electronic Funds Transfer System (EFTS).

EFTS, therefore, may be considered the working prototype of all such planetary data bases, having already reached international size with few rivals either in government or business.[14] We have chosen to illustrate the internal mechanics of the data base with EFTS because no other system is so intensively updated minute by minute. Functionally, as to structure, EFTS erodes the use of barter and paper money through the means of electronically machine-readable transactions, often without the benefit of separate paper verification. The bank, or a consortium of banks, becomes the sole arbiter of your credit reputation. Accumulated information in credit data banks wipes out one's personal privacy and makes cash money more difficult to use than credit itself. Public social desirability is thus reduced to creditworthiness. Until all financial data banks can be linked, no record of total individual indebtedness will be kept; but, the very act of keeping several credit (card) procedures at work keeps an inflationary pressure on the individual and the system (see Fig. 8.2).

The banking industry in both Canada and the United States has been involved in an implosive speedup during the last sixty years as it has passed from mechanical automation to electronic internal accounting procedures. In most large cities, checks can be debited and posted to other branches, indeed, other banks within a twenty-four-hour period. Even though the collection process is somewhat slower in the U.S. than in Canada because of state laws and a more fragmented pattern of banking, it now involves trillions of dollars each day, not merely millions.[15] However, unless the hardware can keep up with general bank business, which shows every sign of increasing in transactional volume, and unless it can instantly present a picture of the depositor's general credit/debit situation involving all his assets, creditworthiness, per se, will become the only index to a person's financial status.

Data Base
(EFTS)

A. Creates cashless
   society

D. Reverses into
   credit-worthiness
   as pure status

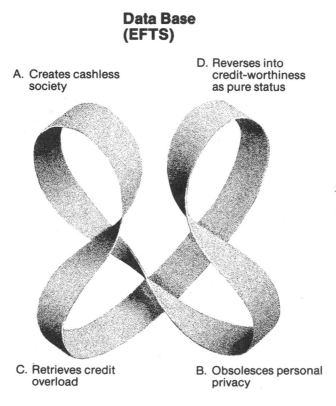

C. Retrieves credit
   overload

B. Obsolesces personal
   privacy

Fig. 8.2.

Since no one will really know the actual adjusted total of any-one else's assets, unless it is attested to by bankruptcy proceed-ings, credit will always be available at the speed of light. One way of understanding this state of affairs is to sketch a future scenario. The year is 1994. You are in a hotel room in New York City. During the night a thief has gained access to your room. He watches your sleeping face as his hands race over your possessions. Then, he locates what he has been looking for—your bank credit cards. Silently, he taps the exit code on the door and lets himself out. Has he taken your cards? No. He has simply copied the ac-count numbers on each of your cards by re-encoding the numbers magnetically onto plastic cards of his own. Within a few minutes he is in the back seat of an automobile, with a portable mini-computer, devising several plastic cards exactly the same size and shape as the ones he found in your luggage.

Later, he travels about the city locating various automatic teller machines (ATMs) which are linked to the various banks (and near banks) with which you do business. At each ATM, he re-checks a list of "broken" personal identification code combinations and punches a few buttons. First he queries the bank computer as to your bank account level and credit limit. Then, the thief electronically withdraws all but fifty dollars (or the daily withdrawal limit). Between $300 and $1000 might glide into his hands from each of your bank accounts. A few days later, when you check out of your hotel, you discover that all your checks have bounced.

As a recent television commercial (1985) asked—what will you do? Fortunately you have kept up your credit card insurance. You dial a Bell Systems 800 number and receive instructions. The person (or computer) at the other end of the connection asks for your bank credit numbers and your current address. You receive an emergency loan for hotel and travel expenses, complete with cash credits. He then issues you a pseudonym, for check-cashing purposes, to be used for the next sixty days while your home bank computer checks with the data-processing centers of all its correspondent banks in the U.S., Canada, and overseas to be sure your account is readjusted for use under your original name. (Of course, it might be better to legally change your name since every sales outlet for goods and services you have been using locally and regionally has received an automatic order shutting off your credit.)

But you have been lucky. Your credit history is long and generally satisfactory and that will ensure that your credit rehabilitation will be relatively short. What is significant about this episode is not that you are cash broke but that you are creditworthy. In other words, unless you are demonstrably without assets, the banks would rather continue lending you money rather than severely limiting your access to credit.[16] The continual rim-spin of the transaction rate will eventually become more important than how much an individual depositor is worth. One of the most identifiable characteristics of the cashless society is that credit is always more important than cash money.

The nexus point in the transactional or payments system is when one passes from the check to the electronic fund transfer, from the left to the right hemisphere, as it were. Despite increased efficiency, it is rarely possible to clear a paper check locally or regionally in less than twenty-four hours. But the computer can debit or credit an amount instantaneously—on an international level, if necessary.[17] That also means that a credit rating can be validated at the same speed. Thus, the electronic passage of credit and money information has created a new service environment.

This new service environment has been constructed from a hybrid merger of the digital computer, automated accounting procedures, and high-speed data transmission on "dedicated" telephone networks. It is broadly called electronic fund transfer (EFT) which one observer has described as "computerized facilities to transfer money electronically between customers' bank accounts." Legal caveats aside, it is really the creation of a super bank through the electric linking of literally hundreds of local and regional data sources to provide the entire Western world a view of your social and economic standing—if your bank transaction is large enough and across state and provincial borders.[18]

Several reasons have been advanced as to why we need a worldwide electronic fund transfer system. The number of actual transactions, whether cash, check, or wire, has become so high that one needs the computer and instantaneous data relay simply to stay ahead. Furthermore, written and printed checks are becoming too expensive to handle—fifty to eighty cents per piece of paper. We have to find a way to reduce the debit float as the number of transactions increases; sometimes in the U.S. it takes three or four days to collect funds as a check passes over state lines and through three or four separate banks. Some experts say it is not so much the transaction volume as the actual value of monetary transfers that is important. Some corporate transfers from several different bank accounts at once can involve hundreds of thousands of dollars. Quick clearance is necessary for safety. Each year in the U.S., corporations transfer among themselves over forty-four trillion dollars; much of this amount is done via electronic transfer.

As the transaction rate increases, fueled by speed-of-light transfer, then information about people's finances will be exchanged at the same pace. At present most banking is done at the branch level. In many major North American cities, however, it is also being done at gas stations, supermarkets, drug stores, and department stores. In other words, a computer terminal is provided at point of sale (POS) and checks are cashed or deposits made by direct connection to a data base, weekends and weekdays on a twenty-four-hour-a-day basis. Within a short time, POS banking will become more common than the teller's cage. And credit via POS will become the major support of all retailing—with the data base decentralized, distributed across all the POS affiliates.

Every merchant has one basic problem; he has to sell goods and services fast enough to achieve a profit in any one sales period. The longer things stay on the shelf, the more expensive they become to the retailer. He can develop sales strategies, such as the "loss leader." He can advertise. The most potent tactic is to offer easy credit. Easy credit is useful as long as the retailer can manage to keep down "no-pay" and fraud losses. To protect himself, he has to maintain a credit assessment operation. Many big sales organizations can afford to have credit departments; most of us are familiar with the fuel company and chain store credit cards. But the small retailer finds credit investigations onerous and expensive. It is at this point that the local and regional banks step in. EFTS has made it possible for the local and regional banks to become the prime credit investigators for all small and medium-sized businesses in a particular geographic area. The instrument which enables the banks to get into credit rating on a large scale is the mainframe computer which can provide instant information not only to the branches but to POS affiliates.

The advantages are two-fold. By building up credit histories on large numbers of potential customers, often from old Credit Bureau files, bank data centers are able to provide useful updated information by phone call. Secondly, the bank, for a fee to retailer and customer, can take the risk out of credit extension. The merchant, therefore, is not obliged to retain his old credit files. Meanwhile, the bank can make up to 21 percent per year on the floating credit balance of the customer. Since most people get paid on Thursday or Friday, the small businessman needs the instant credit line to keep the store open on Monday, Tuesday, and Wednesday. Presumably the ATM will take care of Saturday and Sunday.

The lesson in this arrangement is that for all active purposes the bank becomes the chief arbiter of credit. Since the bank's internal accounting procedures rarely combine accounts on consumer lending and savings nor demand deposit in one centralized file, the data center does not know the day-to-day difference between the customer's overall expenditures and his assets. What it does know is his credit limit, which is roughly based on his transaction record. Each bank the customer deals with, on an independent basis, could provide him with an individual credit limit which may or may not be known to other creditors. And with each individual bank deposit, one might be eligible for a credit card. The opportunities for overextension are limitless.

We now, of course, come to the issue of privacy. Private identity which was tied to a specific time and place is already gone; that is, a definition of self which was achieved in a small community where everybody knew everyone else—the world, as it were, of the nineteenth-century banker. That world began to disappear with the advent of the telegraph.

The kind of privacy we are talking about revolves around the particular information which you can presently hide from a credit investigator. The trade-off for instant credit is no privacy of any kind. Furthermore, the accent on cashless transactions as the transaction rate speeds up will force a great administrative stress on credit investigation. "Trust" will be based on a continual need to update personal information on assets and whereabouts. Eventually, the sheer rate of transactions will force the merger of various EFT credit systems, and this arrangement will encourage personal data trading. High-speed data trading will produce data banks on a national level, rivaling the data bank of the Internal Revenue Service.

The effect will be to decentralize totally the data base. The larger the data-base mosaic, the more difficult, practically speaking, it will be to change bank and branch numbers. At this juncture, a very important result also begins to appear. The user of that credit data base, wherever he is in the world, will have the illusion of centralization as he asks the computer for specific information; but in actuality, he, as well as the data, will be everywhere at once, in the "center" of the system. Time and place in relation to the person will be truly relative.

Once data have reached this state, it is virtually impossible to protect. Anyone with a minimum of expertise can obtain the information. Armed with your account number, outside investigators can now find out your bank balance as easily as a criminal. If they were to acquire dual access codes, maintained by many banks in printed form, they could conduct more sophisticated probes on account balance, available credit, outstanding balance on revolving credit, interest totals, deposit and withdrawal dates, etc.

Credit information for many years has been a tribal business but we have never been totally comfortable with disclosure. The future holds for us a corporate man who will accept the goldfish bowl as a natural habitat—having recognized that electronic espionage has already become an art form.

In that rapidly approaching future, what about the person who has not role-played well enough to attain continual creditworthiness? He will then, as he is now, be a non-person. (And as we all know, non-persons pay cash.) Yet, the definition of status will tend to harden as the EFT payments pattern takes over smaller and smaller transactions so that the debit card will be used for virtually all those payments now assigned to cash. Those who can only pay cash may be looked on as poor credit risks and, consequently, may have difficulty obtaining employment. They will have suffered in effect a high-velocity loss of identity by being tied to hardware in an essentially software environment.

The satellite will, as that specific science matures, complete the process of disengaging man physically and psychically from the earth's surface. Television is figure without ground. A person appearing on regional or national TV is automatically disconnected from his friends, his neighborhood, from the very lifestyle that is his peculiar hallmark. Contrary to his own perceptions, he becomes larger than life and alienated from himself. The same effect occurs with politicians and entertainment personalities, only it is amplified millions of times through repetition. Their personal image becomes frozen into iconic shape. Being a public figure is becoming an archetype, in the same sense that a charismatic figure reminds you of everyone else.

The computer will create large amounts of leisure time for the employed. It will also create for the unemployed on extended welfare time to participate in electronic politics. If the unemployed are also ethnic regionalists, the satellite will body forth new tribal separatists who will make Yasser Arafat seem tame by comparison. In the same way that the Baader-Meinhof Gang was composed of disaffected, unemployed university graduates, every separatist group of the future will have an educated—and therefore skilled—terrorist fringe. Having no fixed place in society, terrorists are discarnate; they have trouble recognizing reality. TV, as a maker of fantasy, reinforces that feeling of disembodiment. A terrorist will kill you to see if you are real. The satellite will distribute terrorist paranoia around the world in living color to match each acceleratingly disruptive event.

Satellites began in 1957 as mere reflecting mechanisms. Today they are radio relays for high-frequency microwaves.[19] Tomorrow satellites will grow beyond the cargo-carrying capacities of the space shuttle and become worlds unto themselves, capable of carrying on high-speed dialogues with earth-based telecommunications machines in excess of anything human beings may understand. The satellite string, or cluster, once in place and safeguarded from sudden disruption, could become a force for decentralization in human affairs which might weaken the written word to the point of dissolution. The satellite surround could replace language as a cultural matrix, using images only as a lingua franca (see Fig. 8.3). What might the satellite surround look like in 1999?

Let us imagine a space shuttle hangs in orbit hundreds of miles above the earth. Its nose is directed toward the stars. Its tail pipes point at an angle toward the mottled blue and brown surface of the planet. An articulated metal arm projects at an angle from the cargo bay. Astronauts are placing satellite parts outside the vehicle. When assembled, and space-taxied into geosynchronous track, the satellite will be enormous, fourteen feet wide and weighing 4000 pounds.[20] Fortunately, space shuttles were enlarged and refitted in the middle 1980s to accommodate heavier loads at higher altitudes; if not, COMSAT officials would have used an advanced model of the Saturn V to send it into geosynchronous orbit, 22,300 miles high.

The time had arrived for the super-satellite with its own power, navigation system, and heavily shielded sensors.[21] Several months earlier, a stripped-down version of a space shuttle orbiter had been on a different mission. The sky had become full of space junk, creating a miasma of radiation interference. This orbiter

Satellite

A. Enlarges global
   information
   exchange

D. Reverses into
   iconic fantasies

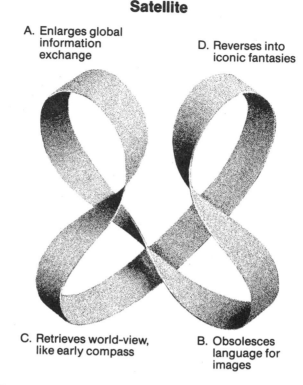

C. Retrieves world-view,
   like early compass

B. Obsolesces
   language for
   images

Fig. 8.3.

carried a single, extremely powerful laser cannon; it destroyed earlier U.S. satellites and foreign space machines in order to create a corridor of clear channels for the super-satellites. Gone would be the "model-T" nuclear reactors and vehicles whose useful life had been only a few years due to high relay failure.

Some of the older satellites had been retrieved, but the remaining machines were blinded or bumped out of orbit to go sailing toward the sun. No mistake about it, this was a hazardous mission because those nations whose satellites were removed were irritated to the point of war. But all concerned recognized that the radio spectrum had been placed in jeopardy by satellites which had drifted into irregular orbit, at the equator.[22]

How did this state of affairs come about? To a certain extent, the space jam was a failure of international regulation. The French, the Germans, the Japanese, acting in private consortiums and selling to the highest bidder, had made it relatively easy for small countries to go into near space. (No one, at any of the space conferences, would agree, for example, on a legal definition of near space.) People all over the world could afford one-meter receiving discs when NHK planar circuits were mass-produced. The Soviet Union finally figured out a way to reproduce computer chip designs it had stolen from the U.S. and made this breakthrough available to most of the Third World countries as a military-diplomatic ploy.

In the United States, the 4/6 gigahertz (GHZ) channels had been selected for the first commercial satellites and soon sporadic episodes of interference occurred amongst earthbound microwave links. One space engineer had conjectured that radio interference really began in earnest when a 4/6 GHZ frequency was chosen by Western countries as an uplink-downlink between earth stations and satellites in an effort to standardize transmissions inside the burgeoning AT&T global satellite network. As satellite communications moved into the late 1980s, lower frequencies around the planet, particularly in mega-metropolises, began to congest. Even though one may regard it as only temporary, land and sea links became overloaded and pressure began to build for bigger, more complex satellites.[23] The bigger the satellite, the less expensive earth stations became. Mass message transmission was literally being lifted off the earth's surface to form a dense electronic skein about the earth.

In the 1990s huge, powerful satellites placed about the planet's girth had more power, more transponders, and operated in higher bandwidths. A much more complex relationship had developed between multiplexing equipment and the number of transponders in use. Back in the late 1970s, commercial earth stations were often massive, having 100-foot dishes and liquid helium–cooled electronics. Now in 1999 transceivers are automatic, small, finely tuned, and portable. Some antennae are no larger than thirty-two inches. The average householder has one attached to his house. The electronic Tower of Babel begins to hum more rapidly.

But let us pause for a moment. Where is this speedup in technology leading? How will people be affected psychologically? First of all, going back to our original analysis of the satellite; it has one prime characteristic—it decentralizes the user, like the telegraph and the telephone. The satellite turns the user into discarnate information. Once placed in relation to the computer/transponder, the user is everywhere at once. You are everywhere and so is everybody else using the system. What is really new about the satellite is that it intensifies the process of being everywhere at once. One can appear simultaneously at every terminal access point on earth or in outer space. This condition poses an almost insuperable problem for the intelligence operative: how can one spy on anyone who is everywhere at once, who could originate anywhere in the net system and change his "location" faster than he can be traced (like Max Headroom)?

The nature of the satellite surround is that it has no center and no margin. "Centers" exist everywhere. This is the way the European understood the character of reality and culture in pre-Renaissance times; no national borders, simply centers of thought and influence; cities which were haunts of being, of ideas—the universe of Duns Scotus and eventually Erasmus where nationalism did not as yet exist. In the age of the super-satellite, large numbers of people will be unable to think merely of regional monopolies of information. Satellites will be able to "talk" to each other and total coverage will lead to total, low-cost communication. Slow information movement will be possible only under the greatest restrictions; espionage will, therefore, begin to disappear.

More and more people will enter the market of information exchange, lose their private identities in the process, but emerge with the ability to interact with any person on the face of the globe. Mass, spontaneous electronic referendums will sweep across continents. The concept of nationalism will fade and regional governments will fall as the political implications of spaceship earth create a world government. The satellite will be used as a prime instrument in a world propaganda war for the hearts and minds of men. The last part of this century will see a war of icons not bombs, a conflict governed by impulse, already begun for us by Roosevelt, Churchill, and Stalin at Yalta.

In the days before the printing press, when the oral tradition still ruled, the values of the medieval world proclaimed that resonance and music were the basis of social order. Then came the Renaissance man and the dogma of *vertu,* the ambivalent Iago espousing competitiveness as a new tempo. The ever-thickening satellite surround will reverse that 400-year development. The shift to individual self-interest and private goals will be played backwards.

The multi-carrier media corporation has the peculiar ability to be a media orchestrator, to link all video-related technologies, whether satellite, earth station, microwave, data base, or computer, into a resonating whole. It is, by its very nature, an affiliated organization, moving any kind of message unit (image, data, or voice) in real time *and* computer time on a speed-of-light network basis.

Because of the diversity and availability of terminal equipment, a number of businesses have joined the older pioneers, AT&T, ITT, and GTE, in the setting up of regional networking. But because of its earlier work with the military and COMSAT, only one telecommunications corporation currently has all the elements to set up global networking, and that is American Telephone and Telegraph (AT&T). AT&T, in cooperation with the new Bell Systems, has a fully developed research arm, Bell Labs, a worldwide manufacturing facility, Western Electric, and free access to all telecommunication links both at home and overseas. Having divested itself of twenty-two operating telephone companies in the U.S. it now has the capital and the investment ability to move worldwide. The management decisions made by AT&T executives in the next fifty years, in association with the Bell Systems (especially AT&T's Long Lines Division), will determine the scope, effectiveness, and size of the first phase of true global networking.[24]

The wired society epitomized historically by telegraph and telephone links has, since the early 1900s, been slowly encapsulated by a wireless canopy of long-distance radio, microwave, and satellite. Coaxial cable has been obsolesced. Open wireless transmission, being truly acoustic, is a group voice. As the satellite surround finally locks into place, software will dictate a shift from left hemisphere to right hemisphere, from the visual to the acoustic—the latter having the prime quality of the interactive mode. In world affairs, decentralization will highlight diversity and fragmentation. But, at the same time, the speed of transmission will

Global Media Networking

A. Instantaneous diverse
media transmission
on global basis

D. Reverses into
loss of specialism;
world-wide synesthesia

C. Brings back Tower of Babel:
Group voice in ether

B. Erodes human ability
to decode in real-time

Fig. 8.4.

inhibit the human ability to decode. As a result, specialism will yield to corporate role-playing on a mass scale (see Fig. 8.4).

Most large corporate organizations have little sense of their social personality. The people at the top who control them only perceive their vastness through numbers; the quantitative not the qualitative. The operative language of most corporate organizations in the Western world is articulated in balance-sheet terms and rarely is the human factor considered. But it is a mistake to expect chief executive officers to act solely in humane ways because the stability of the organization is usually expressed in economic counters if it is a profit-making venture. Yet, it is surprising how an analogy of microcosm/macrocosm can be drawn between the single human person and the corporate organization as a whole. The commercial corporate organization is, after all, a broad extension of the human mind; it develops controlling structures to organize human behavior to produce an economic benefit.

Its activities are analogous, in general, to the interdependent relationship of the left and right hemispheres of the brain.[25] The line functions of a company are like the left hemisphere, concerned with verbal-speech frames of thought which produce quantitative closed-system measurements of the internal and external relationships of the corporate entity. The right hemisphere is concerned with social intelligence which cannot be logically formulated, e.g., the spatial, the musical, the artistic, and the symbolic. The right deals in simultaneous comprehension and the perception of abstract patterns, information frequently provided by staff studies as a preliminary to publicity and advertising campaigns.

Recent split-brain medical research has produced an updated view of the cortex which may have revolutionary consequences for the governance of human society. We have discovered in the last few years that the acoustic dimension, the world of the simultaneous (since we hear from all directions at once), is a sphere whose center is everywhere and whose margin is nowhere. The acoustic world belongs to the right hemisphere of the brain and the left hemisphere is visual—a world of linearity, connectiveness, logic, rationality, analysis, classification, and so on. What is imperative for all of us to know is that at the present time, and for some decades, we have been living in a right-hemisphere world where the major environment in which all Americans and Canadians live is one of instantaneous information. Hence, the right hemisphere of the brain has been covertly present as a ground in all our human relationships; it has been for several generations.

One way of looking at it would be to say that life in the nineteenth century was permeated by the mechanical, the left hemisphere, but that life in the twentieth century, since the advent of electricity, has gradually taken on the coloration of the right. And most people do not know it. In fact, most corporate executives do not know it except, perhaps (and that is a big "perhaps"), those in the business of manipulating electronic technologies.

One American organization, including its Canadian counterpart, does show some cognitive glimmers of its place in the social and psychic fabric: AT&T and the newly divested Bell Systems. A cursory examination of AT&T's public relations publication over the last twenty years shows a growing awareness of the impact of its services on the population at large, for the most part expressed in terms of what would happen if the telephone were precipitously removed. Yet the internal publications of the AT&T company are still doggedly left hemisphere—"don't use the phone, put it in writing." Many Bell Systems operating managers, engineers, and production men seem stonily unaware that they are functioning within an acoustic "miracle," peculiar to a society suffering from a lingering cog-and-sprocket mentality.[26]

The original Bell Systems was from the beginning a right-hemisphere world. When Theodore Vail, the first president, suddenly grasped the positive effects of governmental regulation, he created a structural idea which over a hundred years allowed Bell to grow into a closed, self-sufficient, self-contained unit. It has been holistic in the purest sense. For many years it has made, serviced, and used its own equipment; its profits, by agreement with the government, were primarily based on total corporate assets rather than revenues. It has sought its corporate advantages principally through political rather than economic means; and, in these terms, one can see retrospectively the true message of previous AT&T administrations. That is, the survival of a monopoly form, to ensure a resonatingly "acoustic" organization.

Again, the Bell Systems, still to be taken as a single technical structure today, is essentially acoustic, and simultaneously it is everywhere at once. It has a 360-degree spatial quality; when you place a call or send print data you are immediately in the center of all the systems. The years of laying cable by engineers and production men, like so many spiders building a web, have given a radiating quality to the organization—one breakdown and the tremor is felt in every corner of that seamless electronic web.

However, this structure is undergoing some change.[27] The Federal Communications Commission and various state public service commissions have had something to do with it.[28] Technological pressure from some small, innovative data-processing companies has had an effect. But the main reason that the original Bell Systems has been changed is the clash of two massive technologies, each of which has its corporate advocates: data processing and high-speed information transmission. On one hand stand such giants as IBM, Xerox, and Honeywell; on the other, stands the Bell Systems, long clinging to a near monopoly on high-speed network transmission through AT&T.

How did such a confrontation come about? Most experts trace it to the Carterfone decision in 1968, when the FCC allowed Carterfone Communication Corporation to manufacture and sell telephone instruments and other communication terminals of its own design and, further, to hook up such equipment to the Bell Systems network. Hundreds of companies sprang up to do the same, but in the 1980s, the competition for the terminal business narrowed down to those corporate organizations which could fund and build both computerized terminals and networks.[29] The IBM–Aetna Life and Casualty satellite/microwave networks and the MCI special carrier systems may be precursors of the future.

The purpose of this examination, however, is not so much to write an economic history of the Bell Systems as it is to point out an important social consequence of Bell's current metamorphosis, stemming from the divestiture. One way of looking at the telephone infrastructure is to say that it is the "nervous system" of a wired nation. Like highways and railroads, it has tied us together in a special time-space transmission reference. But unlike highways and railroads, which are sequential in a time frame, the telephone creates a special form of instantaneous contact—it shrinks space to nothingness all at once. To nail the point down, let us study for the moment the social impact of the telephone in its simplest state.

Picture, if you will, a relatively calm day in mid-Atlantic around 1845. Two whalers out of New Bedford are trying to heave to, without reducing speed below a quarter, approaching each other from opposite directions. The first ship is just out from port and the other has been at sea for many months. The captains want to exchange news, so they have their seamen haul them up to the first yardarm amidship. Swinging in seat harnesses, both captains use a small megaphone to shout at each other across several hundred feet of water. The exercise was called the "Nantucket Gam." The skill in such a maneuver lay not only in keeping a somewhat straight course, "by and large" (that is, slightly abaft of the wind), but also in the articulation and vocal strength of the skippers.

The telephone, invented by Alexander Graham Bell reportedly

to improve his wife's hearing, is something like this sailing gambit. It is designed to cut through interference over long distances, eliminate spatial distance, and increase the speed of the human voice. It uses electric technology to do this. However, the very act of using the copper wire as an extension of the human voice produces a peculiar result: it obsolesces the human body as hardware, and in that sense brings back our age-old, right-hemisphere affinity for telepathy. Pushed as a technical process to its limit, a reversal effect occurs. Everyone becomes involved in what was originally meant to be a private communication.

That is, the major social effect of the telephone is to remove the identity of the caller. If he is not identified or chooses not to identify himself, he loses touch with a geographic location and a social function. He becomes truly discarnate and, in that psychic sense, uncontrollable—a phone poltergeist, as it were, who occasionally produces the obscene phone call. Bell Systems, however fragmented, is still one of the nation's largest cooperatives and, therefore, has the capacity to cause the rise of large numbers of acoustic ghosts.

The larger you are, the greater your impact on the social structrue. Actually, when an organization becomes the largest economic grouping in the nation, it *is* the social structure. And for this reason, Bell's organizational changes will produce social mutations of some consequence. It will, of necessity, produce a new kind of tribal man, in the original corporate, or group, sense; the kind of person who has survived the excesses of left-hemisphere thinking and who will have retained his respect for the group awareness of the right hemisphere. But such a mutation will take at least a quarter of a century. He or she will be the psychic precursor of the twenty-first-century ethos, which will be largely lived imaginatively on the planet's surface and in outer space.

AT&T cooperating with the Bell Systems will be instrumental in producing the new man precisely because they are on the cutting edge of the future, due to the hybrid power being released by the merger of the digital computer and high-speed transmission equipment. That explosive encounter, birthing all manner of related technologies, will be reinforced by the fact that AT&T/Bell regionals will still be taken as an associative whole, one of the largest socioeconomic groups on the North American continent, which does not include Bell Canada (767,254 AT&T U.S. employees in 1978; an estimated 315,000 in 1988 after divestiture).

The readjustment battle, which will probably last a generation, will be awesome. A study of the current situation reveals that IBM will probably be the foremost antagonist of the Bell Systems as the technological confrontation moves into the 1990s.[30] IBM will be a lean, wily competitor with an organizational psychology built on salesmanship and rapid equipment turnover, a privilege which only an unregulated corporation could have. The Bell Systems, with a sprawling conglomeration of assets and a ruling concept of maintenance and service (and not marketing), will own something that IBM will try to acquire (and will probably fail to gain)—an exclusive national network of infinite complexity.[31] IBM has been traditionally strong in computer design, especially mainframe and terminals. The strong suit of the Bell Systems has been in long-distance transmission combinations: long lines, microwave, satellite.

In the old days of the West, when the settlers were under attack they would ring the wagons and try to keep the Indians out. If they were greenhorns, they were soon invaded. But as the settlers developed some defense expertise, they invited scouts who knew something of Indian tactics to ride along and help anticipate the redmen's forays. The same thing will happen, and is happening, within the Bell Systems. The AT&T General Department and the former regionally owned and operated telephone companies (OTCs) are hiring executives from the opposition, particularly IBM and Honeywell, to form a defense line against the outside telecommunication rivals. Within a short time, the long-term strategy will shift from defense to aggression as AT&T learns to use the power of its massive corporate assets to relate research and development to marketing techniques. How aggressive AT&T and the Bell Systems can become will depend on heretofore ambivalent FCC policy, which in the beginning mandated AT&T's move away from a noncompetitive atmosphere and into the microelectronic marketplace. As one observer has put it, AT&T is in the difficult position of having to fight with one arm tied behind its back, still hampered by a large corpus of state and federal monopoly regulation.

What will happen to the spun-off Bell Systems internally? First, a great deal of confusion among the middle-level range of managers, whether operational or production-oriented. The average Bell manager, the product of a system of 100 years of right-hemisphere development, is basically a tribal man, taught to maintain an enclosed system and not to be a competitor. The marketing sense will be strange because he or she has been used to responding to calls for service, not finding customers. At the higher echelons, one is more of a mediator, a politician, than a technician. The AT&T General Department in 1979 restructured its administrative group to reflect the new marketing strategy; there are now separate operating divisions for network, business, or commercial customers, and residential phones. Each division has its own sales philosophy. Gradually, each of the twenty-two regional companies (the OTCs), as a result of their legal separation from the General Department, will mirror the head office in each regional area. The strategy specialists, mostly from outside the organization, are essentially "left-hemisphere types."

The mental quality of a person dominated by left-hemisphere thinking is primarily aggressive. The alphabet, from which our form of left-hemisphere thinking springs, produces individualistic thought through its power of abstraction. Using their capacity of categorization and analysis, the left-hemisphere people will reduce the present Bell emphasis on customer service to bottom-line or quantitative results, e.g., "if it doesn't produce a profit, don't do it." The tribal man within the Bell regional OTCs will rebel at first but will gradually be re-educated or eliminated.

In a parallel way, as the regional Bell Systems become more user-oriented, which is the basic touchstone of salesmanship, a great many product lines will develop to fit the "systems design" approach to the customer. Variety in product will cause a new instability at the regional level as each OTC competes to outperform the others. The old hierarchic structure, which depended on loosely related regional organizations, will crack and crumble. AT&T will begin to take on the characteristics of its rivals. All of this will have one root cause—hybridization, the adding to and partial integration of one system with another.

It should be understood that the resultant "child" of hybridization is not the same as either one or the other of its parent systems; for example, the railroad "piggyback," in which a truck body is put on a railroad car. The truck on the railroad car is a hybrid. It does not have the same features as the railroad car or the truck and, hence, is a new entity. As such, it releases new forms of energy. The key to all innovation is a judicious use of parallelism.

Hybridization will force the development of new software, a control mechanism to utilize the new hardware. As more software develops, all hardware will begin to submerge from view, that is, it will become more efficient and breakdown-proof. If IBM keeps ahead of AT&T and Bell Systems in software development, the process of obsolescing the management structure of the old Bell Systems will be accelerated. The rivals will overcome simply by joining AT&T through successful hybridization. In short, after a certain period of instability, AT&T will flip from right-hemisphere dominance to left-hemisphere dominance.

Yet a question arises at this stage. Will the original Bell's customer service be improved by such competition? The answer in the short range is "no." The initial battle of hardware (computer versus rapid data transmission) will end in a merger of ever efficient hardware which by its very durability will cease to gain one's notice, like the can opener and the auto. (The telephone as a single technical instrument has already become invisible.) In the first stage, however, multiple, largely incompatible phone instruments and computerized terminals and allied software will be more vexing than helpful, not to say expensive. In the long run, after the hardware hybridizes and loses visibility, software will finally re-emerge as the key factor. Software, in this case, being the system designed to service the customers' total communication needs in almost an organic fashion.

Now there is one peculiarity about software, particularly in digital computer form, which should receive notice. As it becomes more ubiquitous and easier to use, it intensifies decentralization. Information at the speed of light always tends toward etherealization. At electronic speed all things tend toward an acoustic character and effect. We are now living more in relation to a total environment of electric services (electronic and fiber optic) than our parents were, as they came out of a previous complex of nineteenth-century mechanical services. The new baby Bell Systems have a special sensitivity to this state of affairs.

In fact, Bell Systems managers should use their experience of right-hemisphere thinking to act as a brake or cushion as left-hemisphere ideas begin to speed the exchange of information. As software becomes more important, a strong movement will begin again in the Bell organizations, probably in the next twenty-five years, to return to an emphasis on right-hemisphere orientation—an action-reaction effect, as it were. Remember, as a general rule of thumb, hardware tends toward centralism and connectiveness (left hemisphere) and software tends toward decentralization and discontinuity (right hemisphere). But software will ultimately produce a climate of diversity in the Bell Systems because its final aim is the servicing of the user.

By pushing service and highly individualized software to the limit, you get diversity, which is the essence of right-hemisphere mentality, the realm of the musician and the artist. Of course, as far as AT&T–Bell is concerned, telepathy is the ultimate software. At present, the telephone is still tied to telegraphy and the electro-mechanical environment of the nineteenth century. When AT&T begins to comprehend the full power of parallelism between the use of the long lines/microwave, infrared transmission and the satellite there will be a huge reduction in hardware. If AT&T does not, its rivals will.

The movement toward diversity in communications has, as might be expected, been going on for quite some time. For example, during the studio years (1931 to 1945) the Hollywood film studios were keyed to a mass audience. Everybody went to the movies, including the educated, and film scripts contained, accordingly, a commensurate amount of literacy (e.g., *Gone with the Wind*—1939). But as soon as television appeared to further develop the characteristics of a mass medium, movies became specialized according to audience level. The "art film" suddenly appeared; Disney geared films to the pre-teens, etc. The software, in other words, began to diversify more heavily than ever before. The ultimate in software diversity is the private line. The principle comes into play with television as well. Prime time is designed for the mass audience; but cable is designed to fragmentize mass use. The next step in diversity will not be simply distributive, it will be interactive—a condition in which the user merges with the data base or the system.

If AT&T and the Bell Systems want to survive the competitive war of the eighties and become a worldwide power, they will emphasize diversity on a regional basis as their ultimate weapon. The private user (someone who utilizes special home or commercial information services) will be their prime target. The accumulation of large and sophisticated data bases in the late twentieth century will produce planetary home/commercial high-speed information services utilized by the private user to obtain data for direct personal use. This tailored data will tend to give such a user an illusionary sense of a well-defined identity—assuming the information combination is not available in exactly the same pattern provided to someone else—that's the good news.

The bad news is that all persons, whether or not they understand the processes of computerized high-speed data transmission, will lose their old private identities. What knowledge there is will be available to all. So, in that sense, everybody will be nobody. Everyone will be involved in robotic role-playing including those few elitists who interpret or manage large-scale data patterns and thus control the functions of a speed-of-light society. The more quickly the rate of information exchange speeds up, the more likely we will all merge into a new robotic corporate entity, devoid of true specialism which has been the hallmark of our old private identities. The more information one has to evaluate, the less one knows. Specialism cannot exist at the speed of light.

Angels to Robots:
From Euclidean Space to
Einsteinian Space

BP*   Now that the reader has read the material on the hemispheres and the media, visual and acoustic space,[1] what is the chief thought you'd want him to retain?

MM   That he's not living in a natural environment. If he's civilized, he's living in Euclidean space—closed, controlled, linear, static—abstracted from the world around him. Like language, it is an attempt to manipulate as well as interpret the world.

BP   Something like a baby in its crib. Have you noticed how the small child will try to put everything—dishes, rattles, balls—into its mouth?

MM   Yes, he's trying to "swallow" his world by taking it into himself—to control it in other words. Have you noticed that one cannot visualize geometric figures except in a void? This characteristic is an essential clue to understanding Euclidean space. It is not the whole of nature, it is an abstraction, an imaginative invention. For over 2500 years,

* Dialogue between the authors, 1978.

the concept has so conditioned our thinking that we are virtually forced to live by cubes and rectangles—square rooms and houses, matching streets. We can't feel comfortable with a circle in architecture unless we've squared it. The Euclidean straight line or plane has taken over our brain, or at least part of it—the left hemisphere.

BP   But how did it all start?

MM   With the alphabet. The Phoenicians brought it west to Greece, probably from the Orient. The Phoenicians exported the ideas of one to nine as well, but as you know numerical manipulation, which has its roots in the alphabet, didn't work too well until the concept of zero freed it up. Look at our alphabet; it has the four simultaneous aspects of a square. It is, in each one of the alphabetical signs, continuous, connected, homogenous, and static. The characteristics also of visual space. A succession of fragmented bits having no real meaning except what we read into them. The best way to understand the essence of the alphabet is its progressive movement of "one-thing-at-a-time."

BP   Then, the most obvious question seems to be if the alphabet gave rise to Euclidean space and Euclidean space is not a complete way to visualize the totality of the world, how does one imagine nature?

MM   Well, let's first deal with the word "nature." It is itself an abstraction. We know that literate Greeks had to find a word for their ability to abstract visual order out of the environment which surrounded them. So, having identified a number of balances or cause and effect equilibriums, they called them nature (*physis*) and everything else was chaos. And you find this idea being adopted right down the time tunnel from Parmenides to Descartes, Galileo, Hobbes, and Locke.

BP   And Milton, Nietzsche, Heidegger.

MM   And a cast of thousands! Man's view of his environment was not percept but concept—an almost total extension of one sense, his eye. And a sense keyed to the horizon, to a proportionate sense related to the vanishing point. He has used his eye to create a squared-off, controlled environment, whether he is at home or at work, on the highway, in a train or a plane, and everything outside is "nature"— which by now he has identified with chaos.

BP   Wait a minute. Let's back up. Did you say that modern man's definition of nature is not the same as that of the ancient Greeks?

MM   There is a subtlety involved here. The Euclidean world or the mechanical world has become so familiar and comfortable to modern Western man that he views everything outside his windows as wild. "Let's go into the wilderness for a hike," he'll say to his kids—meaning nature. A low-level transference of meaning.

BP   Can modern Western man really experience the wilderness?

MM   Not really. Because his senses are imbalanced.

BP   When you use the term senses, you're really using it in the older meaning of sensorium: vision, hearing, touch, taste, and smell? Aristotle's use of that term in Greek which found its way into the Latin as such.

MM   Yes, but perhaps with a difference. I'm more inclined to use it as a variation of the Lucretian frame of reference. Lucretius was inclined to think that there were four senses, with vision being a form of touch. I'm disposed to agree with some modern researchers who recognize four major divisions of the senses but taste being a variant of smell. That is, four senses: vision, hearing, touch, and smell. All technology is an extension of these four capabilities.

BP   Modern man's world is a mechanical world because it is based on the straight line or plane.

MM   Yet a continuum does not exist in nature, or better still, the
      total environment, the material universe. There are no
      straight lines in space; as Einstein pointed out parallel
      lines do not meet in infinity. They simply curve back upon
      themselves. We have invented the straight line to give us a
      sense of location on the Earth's surface. But Euclidean ref-
      erence will not work in outer space. True nature, as we
      should understand it, is acoustic. Acoustic space has no
      center. It consists of boundless random resonations. It is
      the kind of orientation we have when we are swimming or
      riding a bicycle—multisensuous, full of kinetic spaces. Eu-
      clidean mathematics has not a real grasp of the acoustic; it
      is too rational. Boolean concepts of algebraic possibilities
      might, however, be a place to start. . . .

BP   When you say acoustic you mean the experience of re-
      turned sound?

MM   Yes, echo. Acoustic is ECO-LAND. And our cortex is
      divided into two hemispheres, one of which concerns it-
      self with visual or Euclidean space and the other with acous-
      tic space.

BP   Nevertheless, modern man does not have equal access to
      both hemispheres in apprehending the world around him.
      One of the hemispheres, the left, is dominant in Western
      culture and the other seems to be operating at a minimum
      level.

MM   Perhaps a better way of looking at it is to say that we are
      engaged in the Herculean task of processing right-hemisphere
      perceptions of the world through the left hemisphere. That
      is, everything we experience in some way has to have a logi-
      cal cause and effect relationship or we are unhappy.

BP   We are constantly suppressing the awareness that the mate-
      rial universe is comprised of resonances; that no straight
      lines exist.

MM   Exactly. Because the Euclidean construct is controllable. The "center" of acoustic space is everywhere and, therefore, seemingly chaotic.

BP   Acoustic space is hard to imagine.

MM   Well, start with deep space. Visualize yourself looking out the cockpit window of a spaceship moving into the area beyond the immediate galaxy. There are no recognizable planets or stars, just pinpoints of light and swirling gases rushing toward you, streaming away on either side of your field of vision.

BP   Like the opening sequence of the television series, *Star Trek,* or the scene in *Star Wars* when Hans Solo puts his interstellar freighter into warp speed as they move across the galaxy to find the rebel planet.

MM   Precisely. But the *Star Wars* scene is somewhat more accurate because at the exact moment Solo puts his spaceship into warp speed all those pinpoints of light stand still. He's traveling faster than the speed of light and thus the space freighter becomes simultaneous and everywhere at once— the properties of acoustic space.

BP   Does it surprise you that those Hollywood special effects men would anticipate your description of acoustic space?

MM   No, because motion pictures are a blend of Euclidean and acoustic thinking, of the mechanical and the electrical. As long as they were confined to imagining Earth spaces they would lean toward the Euclidean. But, as soon as they had to imagine the consequences of traveling faster than the speed of light, then they translated themselves into an acoustic environment. The imagination is most creative in acoustic space. Acoustic space has the basic character of a sphere whose focus or "center" is simultaneously everywhere and whose margin is nowhere. A proper place for the birth of metamorphosis.

BP   Let's review for the moment. One of your major ideas is that ever since the transition was made from the tribal to the literate society in the Western world we have been developing an ever increasing complex of Euclidean thought which has permeated our entire way of life. Euclidean thinking is a way of retreating from all of the material universe to create a small manageable environment which one can ultimately control. It is per se an abstraction and characterizes all human artifacts, from the baby crib to the windowless skyscraper.

MM   Euclidean thought emphasizes the mechanical and is focused on the creation of hardware. It is basically centralist in its tendencies. Remember, visual space as elucidated in Euclidean geometry has the basic characters of lineality, connectedness, homogeneity, and stasis. All Western societies, in common, strive toward equilibrium or stability—frequently one society seeking stability at the expense of another.

BP   And the development of literacy or the refinement of the alphabet generates Euclidean control; and, the last 2500 years from the pre-Socratics to the present day is really the history of ever growing centralism.

MM   Yes, but don't forget, none of this was foreseen. When Euclid (365?–300 B.C.) wrote the *Elements* it is unlikely that he conceived of himself as striking out on a new level of consciousness, despite his training by the successors of Plato. He was a teacher in Alexandria around the time of Ptolemy I and probably was working furiously only to make himself useful to the Egyptian priest-architects.

BP   Yet he must have had some inkling of the closed-system kind of thought he was refining. The last few books of the *Elements* deal voluminously with irrational numbers.

MM   Euclid and the Egyptians must have gotten on famously. After all, the pyramid is truly an unearthly triumph—a totally self-contained world placed in a void of sand.

BP By that measure, the Egyptians were among the first civilized men.

MM We should, of course, use your thought as a touchstone. To be "uncivilized" is to be uncentralized. Civilization is Euclidean. Primitive society is acoustic and oral. The oral world is primordial. It responds to the simultaneous, the holistic, the harmonious—it is literally the abode of song—for us, of Runes and Nordic chants, the keening of the Scop in *Beowulf*.

BP *Beowulf* brings us to the entryway of the Middle Ages and in itself is a bridge between the oral society of the Angles, Saxons, and Jutes and the Euclidean influence of literate Christian peoples. What is important about the story of Beowulf and Grendel is what the transcribing monks left out.

MM The Christian church by A.D. 900 was thoroughly Romanized, as was to be expected. St. Paul was a Roman and gifted the early communities with his background in Roman law and administration. And what was that mechanism of law and administration based upon?

BP Paper, or more exactly papyrus.

MM That's right. Engraving in stone is for the priests; they have an affinity for spanning eras. But soldiers are no-nonsense managers. They need to deal with the here and now. The alphabet and paper create armies, or rather the bureaucracies which run armies. Paper creates self-contained kingdoms at a distance.

BP I remember one historian saying that when the silver mines and the papyrus began to run out, the Empire began to decline. About the period of Marcus Aurelius.

MM The transport of messages across the Roman Empire created a huge network of roads and naval passages on the seaways. And all this involved a speedup of information.

The basic function of the Roman soldier was to keep the messages going out and the tribute coming in. An initial speedup of information promotes centralism and an emphasis on the material.

One way of thinking about the era of the pre-literate is that it is a period of emphasis on group or tribal identity. Ownership is in common. The tribal person cannot think of himself in any other way but as a member of the group. This is the dramatic world of Aeschylus. Sophocles and Euripides chronicle the destructive rise of the individualist. Petronius's *Satyricon* simply tells us how the Romans carried the concept of individuality to ridiculous lengths.

BP   Wasn't individualism lost in the Dark Ages? Wasn't there an interruption of the forward, cumulative thrust of Euclidean thinking?

MM   Not really. The hand of Marcus Aurelius passed the bay leaves of individuality to the Christian church. It was the true successor to the Roman sense of hierarchy. The early church fathers discovered one very disconcerting fact: tribal man is not very educable. Convince him to one point of view one day and he goes back to his old ways the next. The pre-literate has a short memory. But literate gentiles were more receptive to change. The gospel of good news was spread by literates who had cultivated the habit of putting their memories on paper. The history of Europe between A.D. 300 and A.D. 950 was a story of the task of missionizing the barbarians into literacy.

BP   And reproducing in another key the Roman way of life. The medieval manor was really an update of the Roman villa.

MM   Notice, however, that the movement of information was a restraining factor. The medieval kingdom seldom stretched beyond the horizon. Putting the manor or the castle keep on a hill may have extended surveillance room but most early medieval kingdoms were circumscribed by how much territory a group of armed men could traverse in a day's ride. Riding bareback on a blanket was none too steady and very uncomfortable.

BP Hence, Europe was a patchwork quilt of tiny kingdoms, constantly squabbling. What changed that?

MM The stirrup. The secret weapon of the marauding Hun. It enabled a man to stand in his saddle and to exert a downward pressure on the horse's back as he swung his weapon. The knight became an invincible tank. Large-scale conquest followed. Consolidation began and the speedup of information gained momentum once again. Admittedly, the great plagues of the twelfth and thirteenth centuries interrupted this process for a period of time. But since the introduction of the alphabet, the concept of centralism had gradually gathered force until it reached a complex apotheosis in the nineteenth century.

BP You are talking, I take it, about the process of metamorphosis, or chiasmus. Every process pushed far enough tends to reverse or flip suddenly. Surely this condition must have begun earlier than the nineteenth century.

MM It actually started with the Greeks before Christ, when old men used to amuse children by rubbing amber against cloth to make static electricity. Electricity has all the properties of the acoustic world: it is simultaneous and everywhere at once. A 360-degree quality. Man found it difficult to understand electricity until he could contain it. Without a constant, controlled source of electricity, one can hardly hope to fathom its uses. One experiment after another failed until the invention of the Leyden jar in the eighteenth century, the possibilities of conduction with Galvani, the Voltaic pile of Alessandro Volta in 1800. Michael Faraday crowned the search with the development of rudimentary electric motors and generators by 1821. It was not until Edison in 1868 that many practical uses of electricity emerged. It took 3000 years to harness the force which is the opposite of Euclidean centralism—electricity and its acoustic properties.

BP   I agree with you. Thomas Alva Edison was truly the first great experimenter of electrical uses, with due respect to Michael Faraday and Nikola Tesla. Yet that which made Edison great was not his incessant curiosity but his vow after 1868 to invent only for actual use.[2] As you know, his first invention, an electrically driven stock ticker, was a mechanical success but sold poorly.

MM   He decided then to assess the market by percept instead of prior assumptions.

BP   He concentrated on those telegraphy-related machines which would speed up the rate of information transfer. In fact, the idea for the phonograph came to him as he listened to the irregular whine of a high-speed telegraph tape as it raced through a repeater. He adapted a telegraph arm with a telephone diaphragm to reproduce a recording of his own voice.

MM   A flip point from the mechanical to the acoustic—the rounded, tin-foil cylinder. New symbol of the coming acoustic age.

BP   Marshall, talking about the pattern of chiasmus, or metamorphosis, over a span of 2500 to 3000 years does give it the "unreal" aura of the mythic. It appears to happen over too long a span of time for mere mortals to grasp. Surely the pattern is recognizable in shorter time periods and more mundane instances?

MM   Of course it is . . . at all levels of experience. But before discussing such examples, I must repeat once again that the chiasmus pattern is really a play of metaphors (tetrad), that the intensification of any human process, artifact, or creation will have four simultaneous consequences: it will enhance something, it will cause something to become obsolescent, bring something back, and pushed to the limit flip into an opposite effect. As a point of percept, the introduction of money enhances the rate of exchange and causes barter to become obsolescent. As exchange increases, the phenomenon of potlatch occurs, and finally money flips into nonmoney or the credit card. Take the telephone, it enhances the speed of the human voice; it causes the body as hardware to become obsolescent, retrieves telepathy (ESP, mysticism, the occult), and at last flips into the group voice or omnipresence.

BP   The group voice or omnipresence?

MM   Oh, you know, the conference call. The echo effect one hears from the phone speaker is really the sound symbol of the flip point, or chiasmus. The telephone in its larger context suggests a slightly broader area to explore, one which you and I have been examining for the last few months (1978).

BP   You mean, Bell Systems and AT&T? The present state of the Bell Systems Company and its parent company, AT&T could illustrate the chiasmus pattern?

MM   A case in point. About seventy-one years ago Theodore Vail, the first president of AT&T, recognized that government regulation was inevitable. So he accepted the condition of state (and eventually federal) regulation in exchange for a monopoly situation. That is, a profit structure tied to the amount of corporate assets rather than the marketplace. Such a monopoly situation was most necessary because the telephone industry was embarking on the huge task of wiring the nation.

BP   Yet, state and federal regulators have been intermittently fearful that the monopoly situation would result in a Bell/ AT&T takeover. After all 50 percent of all U.S. phones were installed by Bell companies in 1913. By 1979, that figure was 82 percent.[3]

MM   So, in 1913, and again in 1949 and 1978, the Justice Department initiated divestiture actions. Nevertheless, so far, the basic network structure has remained intact. And what is that structure: basically acoustic and right hemisphere.[4] It has the intrinsic nature of a sphere, simultaneously resonating, whose center is everywhere at once.

BP   The regional Bell companies are all plugged into one another through the Long Lines Division. Historically, it has been a loose confederation of twenty-two different companies, organized regionally, with common interests. Since it has been a monopoly so long, the chain of command has been, up until recently, very direct. The president of the General Department (AT&T headquarters) in New York City would deal on a personal level with each of the heads of the regional companies, the OTCs. The development, manufacture, and re-cycling of equipment would be taken care of by wholly owned subsidiaries. All pretty self-contained.

MM   Another way of saying that it is holistic. Like every acoustic environment it has to be self-contained. Then what happened?

BP   It would appear that Bell personnel were primarily tuned to the job of taking orders and servicing prior requests from the community at large. It wasn't used to a marketing condition where it had to compete with other companies to service segments of that community. The complex of Bell companies was not innovative in either equipment or service, as some federal regulators maintained. In consequence, the FCC allowed Carterfone, a small manufacturer of switchboards and terminals, to attach its equipment to Bell telephone lines. And a race began to set up a new industry which linked the digital computer with high-speed data transmission devices.

MM   The Bell companies and AT&T did not respond at first?

BP   Not at first. Then it became clear that very large competitors, like IBM and Honeywell, were shoving the smaller companies aside for the "hookup" business and a very strong vertical managerial structure began to show itself in the Bell organization. AT&T in 1978 set up three new divisions at its New York headquarters, according to the shape of its customer revenues: network, business, and private residences. Network is basically long-distance connections; business means the business corporations serviced by Bell; and private residences means home phones. Each area is administered separately and mirror structures of these divisions have appeared in all twenty-two regional OTCs.

MM   Bell/AT&T is obviously afraid that IBM, its closest competitor in the business area, is going to set up another national telephone company, if government deregulation permits it.

BP   And, if given state and federal clearance, IBM could. In fact, it has already set up a satellite corporation and that piece of hardware, plus microwave connections might make it possible. Since IBM is superior in marketing computer systems, it might gain a serious competitive edge over AT&T.

MM   What we are talking about once again is a chiasmus pattern. Bell/AT&T extended its acoustic character so intensively that there was a flip to its opposite side. The Bell companies are on their way to an energetic left-hemisphere experience, with a collateral shakeout of personnel and refinement of equipment. The primary area of growth will probably be in the business area in computerized data transmission—that will enhance the instantaneous sending of large amounts of data, yet at the same time, it will obsolesce meaning (the human ability to decode). It will retrieve pattern recognition (of a party-line variety) but flip into a loss of meaning, which is a loss of identity. The same thing will happen with EFTS. It was begun as a way to improve the banking industry's ability to keep a heavy cash flow going through the community. Strong, consistent spending during every business day is necessary to keep the economy from stalling. But in such a setup, the bank computer validates a person's creditworthiness; all other satellite credit operations, in department stores, food chains, gas companies, and hotels, begin to eliminate their credit checks and depend solely on regional banks. As the speed of credit data exchange picks up, the bank's determination of a person's credit rating becomes the paramount ranking. The accent on cashless transactions, instead of cash money, forces a great managerial stress on credit investigations. Mergers of EFTS regional systems will encourage credit information trading, central data banks, and regular systematic probes of people's personal lives. Privacy disappears. Again, the transmission of data at the speed of light creates nonpersons.

BP   Is there any use of the digital computer which could create a new bulwark for the individual?

MM   Yes, what some researchers have called the new home information services in which the computer is utilized by a person to organize particular data needs; that is, to order groceries or machine parts, home security, specialized news items, answering services, and paid work at home. A computer as a research and communication instrument could enhance retrieval, obsolesce mass library organization, retrieve the individual's encyclopedic function and flip into a private line to speedily tailored data of a salable kind.

BP   Well, in what way could such a service create a new kind of personal protection?

MM   It could create more personal leisure. Having more leisure will encourage people to "drop out" to enhance their sense of identity.[5] If their jobs are becoming routine and not helping the necessities of self-definition, then being able to get that occupation done within less time than was previously required, will enable people to set aside large blocks of leisure time with which to explore a hobby, a sport, a secret avocation. In other words, more time to "drop out" and "tune in" on themselves.

THE UNITED STATES
AND CANADA: THE BORDER
AS A RESONATING INTERVAL

Epilogue: Canada as Counter-Environment

Canadians and Americans share something very precious: a sense of the last frontier. The Canadian North has replaced the American West. That primeval woodland, that vast wilderness is there, from Banff to New Foundland, giving all North Americans a spatial habitation Europeans do not know. For two centuries, at least, the frontier has taught us how to go out alone.

As it did for nineteenth-century plainsmen, going out to be alone raises the ultimate question: who am I? We remove ourselves from the anonymity of the crowd. Standing on the edge of the Grand Canyon or a glacial tundra, we are swept with a sense of immensity, a feeling of awe, which—for most of us—is swiftly followed by a prayer and thanksgiving. For those breaking the Oregon Trail, the wilderness was a red-clawed menace. Today we reclaim and repossess ourselves in forest and glen and take stock, once again, of our individual worth.[1]

We have tried to demonstrate how video-related technologies, taking advantage of left-hemisphere overload, will implode our inner sensibilities, destroying a previous imbalance between the hemispheres. But more than that, these technologies will invade our inner peace, occupying our every waking moment. We will need a place to hide.

Earlier in the century, Charlie Chaplin seemed to find a partial answer. He was the overburdened European turned inside out. In the time of Dickens, English coketowns allowed no man to be idle. But as soon as the immigrant left Ellis Island, he discovered a marvelous thing about America: he was free to do anything and go anywhere he liked. He might have been Huck Finn exploring the mighty Mississippi. You will remember that Huck told Tom Sawyer that loafing, carefully pushed to its limits, defined the meaning of independence. As he and the slave Jim floated down the river they could carelessly lose their goals and objectives. Loafing could reveal your true self because you are "dropping out" to be alone. Chaplin, like Huck, epitomized the transformation of loafing. Spiffy, elegant in bowler, cane, and morning coat, he became a footloose knight of the road.

Carefree individuality, though, is foreign to the European. He doesn't find privacy in the great outdoors but in the crowd itself. He spends his lifetime learning the strategies and uses of the social mask. Like the legendary Heidelberg prince, he must wear a tribal and corporate mask in the same way he does his uniform. Fixed economic and social rank obsolesces nonchalance from the moment of birth. And release can only be obtained briefly in a *Fasching* or Mardi Gras. The European habitually goes out to be social and comes home to be alone. The American and Canadian do exactly the reverse.

But electronic technologies have begun to shake the distinction between inner and outer space, by blurring the difference between being here or there. The first hint of this condition came with the telephone. By increasing the speed of the private voice, it retrieved telepathy and gave everyone the feeling of being everywhere at once. After teleconferencing is established, the picturephone will be reintroduced, taking the user outside for public inspection whether he or she is ready or not.

As the border is gradually erased between inner and outer space, between the aggressive extroversion of the marketplace and the easy sociability of the home, North Americans will need another refuge, a place where nostalgia, for example, could serve as a link with the stability of times gone by. If a U.S. citizen so chose, Canada could become an enormous psychic theme park; something like a Hollywood set that simultaneously links the past with the present, the city with the wilderness. The Province of Quebec seems to have anticipated this role with its recent advertising slogan, "Foreign yet Near." The calculated ambivalence of the Canadian is a most efficient way of maintaining a low profile, as a receptive ground for other people's fantasies.

In the nineteenth century the Germans were expert at the art of discovery, that is, deciding what effect they wanted and working back step-by-step to uncover the starting point of the thing to be discovered. In a similar way, the Canadians are masters of what Bertrand Russell has called the twentieth century's highest achievement: the technique of suspended judgment. Canadians experiment with technology from all over the world, but rarely adopt any technical stratagem broadly. For example, in their current examinations of teletext, Canadian telephonic engineers will test both U.S. *Telidon* and British *Ceefax,* with a side nod to Gallic *Antiope,* while resisting AT&T's efforts to standardize all teletext transmission equipment.

Canadians are always waiting for the latest model without making a commitment to what is available here and now. As the United States careens toward its rendezvous with the unified effects of combined video technologies, it might steadily keep its eyes on the rear-view mirror—as indeed all other previous cultures have done in terms of the introduction of new technical artifacts—to see how the Canadians sidestep the impact of these new media, keeping a sort of stasis in place so characteristic of the northern ability to juggle fierce separatism and regionalisms without cataclysmic finality.

A border is not a connection but an interval of resonance, and such gaps abound in the land of the DEW line. The DEW line itself (Distant Early Warning system), installed by the United States in the Canadian North to keep this continent aware of activities in Russia, points up a major Canadian role in the twentieth century—the role of hidden ground for big powers. Since the United States has become a world environment, Canada has become the anti-environment that renders the United States more acceptable and intelligible to many small countries of the world; anti-environments are indispensable for making an environment understandable.

Canada has no goals or directions, yet shares so much of the American character and experience that the role of dialogue and liaison has become entirely natural to Canadians wherever they are. Sharing the American way without commitment to American goals or responsibilities, makes the Canadian intellectually detached and observant as an interpreter of the American destiny.

In the age of the electronic information environment the big nations of the First World are losing both their identities and goals. France, Germany, England, and the United States are nations whose identities and goals were shaped by the rise of the self-regulating markets of the nineteenth century, markets whose quantitative equilibrium has been obsolesced by the dominance of the new world of instant information. As software information becomes the prime factor in politics and industry, the First World inevitably is minus the situation which has given meaning and relevance to its drive for mere quantity. New images of identity based on quality of life are forming in a world where suddenly *small is beautiful* and centralism is felt to be a disease.

In this new world the decentralized and soft-focus image of the flexible Canadian identity appears to great advantage. Canadians, who never got "delivery" on their first national identity image in the nineteenth century, are the people who learned how to live without the bold accents of the national "egotrippers" of other lands. Today they are even more suited to the Third World tone and temper as the Third World takes over the abandoned goals of the First World. Sharing many characteristics of the Third World, Canada mediates easily between the First and Third Worlds.

No one better understood the advantages of being nationally a "nobody" than George Bernard Shaw whose borderline frontiersmanship created his great career on the stage. In the preface to *John Bull's Other Island,* Shaw explained:

> When I say that I am an Irishman I mean that I was born in Ireland, and that my native language is the English of Swift and not the unspeakable jargon of the mid-XIX century London newspapers. My extraction is the extraction of most Englishmen: that is, I have no trace in me of the commercially imported North Spanish strain which passes for aboriginal Irish: I am a genuine typical Irishman of the Danish, Norman, Cromwellian, and (of course) Scotch invasions. I am violently and arrogantly Protestant by family tradition; but let no English Government therefore count on my allegiance: I am English enough to be an inveterate

Republican and Home Ruler. It is true that one of my grand-fathers was an Orangeman; but then his sister was an abbess; and his uncle, I am proud to say, was hanged as a rebel. When I look around me on the hybrid cosmopolitans, slum poisoned or square pampered, who call themselves Englishmen today, and see them bullied by the Irish Protestant garrison as no Bengalee now lets himself be bullied by an Englishman; when I see the Irishman everywhere standing clearheaded, sane, hardly callous to the boy-ish sentimentalities, susceptibilities, and credulities that make the Englishman the dupe of every charlatan and the idolater of every numbskull, I perceive that Ireland is the only spot on earth which still produces the ideal Englishman of history.[2]

Like Shaw, the Canadian "nobody" can have the best of two worlds—on the one hand, the human scale of the small country, and on the other hand, the immediate advantages of proximity to massive power. Knowing the United States like the back of his hand, the Canadian can be playful in discussing America. He is happy to invite "the ugly American" to enjoy the idyllic play-grounds of a largely unoccupied land of lakes and forests, whether of Quebec or Ontario or British Columbia.

If there are 250,000 unnamed lakes in Ontario alone, there is an even larger problem of toponymy in tracing the Canadian lan-guage. Morton Bloomfield, a professor of English at Harvard Uni-versity and a Canadian by birth, surfaced one of the many hidden borderlines that interlace the Canadian psyche when he explored the character of Canadian English, a subject neglected by both Ca-nadian and American scholars. In "Canadian English and Its Re-lation to Eighteenth-Century American Speech," he noted:

The probable explanation for this neglect lies in the fact that most American investigators, ignorant of Canadian history, are under the impression that Canadian English, as undoubtedly is the case with Australian, South African, and Newfoundland English, is a direct offshoot of British English and therefore does not belong to their field of inquiry. It is, however, necessary to know the history of a country before one can know the history of its language.[3]

Bloomfield pointed out that in *The American Language,* H. L. Mencken shared the widespread illusion that American English conquered the British English of Canada, whereas Canadian English had been American from the time of the American Revolution:

> After 1776, however, the situation changed and a large increase in population occurred, entirely owing to the movement north of many Tories or Loyalists who wished, or were forced, to leave the United States because of the American Revolutionary War. They carried with them, as a matter of course, the language spoken in the Thirteen Colonies at the time.[4]

Without any self-consciousness English Canadians enjoy the advantages of a dual language. Canada is linguistically in the same relation to the United States as America is to England. Stephen Leacock humorously varied the theme:

> In Canada we have enough to do keeping up with the two spoken languages without trying to invent slang, so we just go right ahead and use English for literature, Scotch for sermons and American for conversation.[5]

Another psychological borderline shared by Canadians and Americans is a legacy of their nineteenth-century war on the empty wilderness, as indicated in Lord Durham's *Report on the Affairs of British North America* (1839):

> The provision which in Europe, the State makes for the protection of its citizens against foreign enemies, is in America required for what a French writer has beautifully and accurately called, the "war with the wilderness." The defence of an important fortress, or the maintenance of a sufficient army or navy in exposed posts, is not more a matter of common concern to the European, than is the construction of the great communications to the American settler; and the State, very naturally, takes on itself the making of the works, which are matters of concern to all alike.[6]

It would be strange indeed if the population of North America had not developed characteristic attitudes to the spaces experienced here. A century of war on the wilderness made customary the habit of going outside to confront and explore the wilderness and of going inside to be social and secure. Going outside involved energy and effort and struggle in frontier conditions that called for initiative amidst solitude. Thus, Margaret Atwood notes in her critical study *Survival:*

The war against Nature assumed that Nature was hostile to begin with; man could fight and lose, or he could fight and win. If he won he would be rewarded: he could conquer and enslave Nature, and, in practical terms, exploit her resources.[7]

Atwood's study of Canadian writers reveals a frontier trauma, yet one that is not uniquely Canadian. Twain's *The Adventures of Huckleberry Finn,* Whitman's *Leaves of Grass,* Thoreau's *Walden,* and Melville's *Moby-Dick* record new attitudes to both inner and outer space; spaces that had to be explored rather than inhabited. Here, then, is the immediate effect of continental space: to seem to be a land that has been explored but never lived in. The Oriental comment is not without good grounds: "You Westerners are always getting ready to live!" Expressed in Whitman's "Song of the Broad-Axe" and C. C. Moore's " 'Twas the Night Before Christmas" ("The stockings were hung by the chimney with care") lie the two psychic poles of the special North American feeling for space—the outer space for aggressive extroversion and the inner space for cozy sociability and security amidst dangers. On the borderline between these areas of aggression and hospitality Hawthorne and Henry James etched their psychic adventures and "the complex fate" of being a North American. "It's a complex fate, being an American," James wrote, "and one of the responsibilities it entails is fighting against a superstitious valuation of Europe." To become cultured in America while resisting European values became a major theme in Hawthorne and James. Neither thought to consider the hidden physical polarities of this continuing conflict.

The mutual feeling for space in Canada and the United States is totally different from that of any other part of the world. In England or France or India people go outside to be social and go inside to be private or alone. By contrast, even at picnics and camping holidays and barbecues North Americans carry the frontier with them, just as their cars, their most cherished form of privacy, are designed for special effects of quiet enclosure. Where a European thinks of "a room of one's own," the North American depends upon the car to provide the private space for work and thought. Typically, one can see *through* an American car when driving, but one cannot see *into* the car when standing. The reverse is true of a European car; one can see into it but not through it when on the road.

Since we are seeking to delineate some of the Canadian border-lines, it is natural that those psychologically shared with the United States are the areas of maximal interplay and subtle interpenetration. One mythic borderline Canada shares with the United States springs from the heroic deeds of Paul Bunyan and Babe, the Big Blue Ox. It is generally accepted that Paul Bunyan was an American logger of the 1840s and 1850s. The folk art of the tall story is dear to the frontier, that world of the resonant interval where public amplification proliferates. These tall stories are often called in to calm the ardor of those who delight in exaggeration. The Paul Bunyan man can retort with a story of the huge pines in his territory which he began to notch with Paul. After notching for an hour or more, they went round the tree and found two Irishmen who had been chopping at the same tree for three years, including Sundays. Paul is a frontier or borderline figure who is a continent-striding image. Newfoundland poet E. J. Pratt had a special gift for the gigantic in verse as in his *Witches' Brew* and *The Cachalot*. The frontier poet or novelist will feel "the call of the wild" rather than the lure of the parlor or even the pub. As Thoreau wrote in *Walden:* "I have never found the companion so companionable as solitude."

The frontier is naturally an abrasive and rebarbative area which generates grievance, the formula for humor. Thus the first major Canadian literary character was Sam Slick, the Yankee clock-maker. The frontier abounds in figures of fun-writing like Stephen Leacock, Canada's Mark Twain.

Hugh Kenner looked into the borderline matter in his essay "The Case of the Missing Face."[8] He begins his quest for the missing face of Canadian culture with an observation of Chester Duncan: "Our well-known Canadian laconicism is not always concealed wisdom, but a kind of . . . between-ness. We are continually on the verge of something but we don't quite get there. We haven't discovered what we are or where we're going and therefore we haven't much to say." Duncan found the key with "between-ness," the world of the interval, the borderline, the interface of worlds and situations. It may well be that Canadians misconceive their role and opportunities and feel the misguided urge to follow the trendy ways of those less fortunately placed. The interface is where the action is. No need to move or follow, but only to tune the perceptions on the spot.

Harold Innis, the Canadian pioneer historian of economics and communication, imaginatively used the interface, or borderline situation, to present a new world of economic and cultural change by studying the interplay between man's artifacts and the environments created by old and new technologies. By investigating social effects as contours of changing technology, Innis did what Plato and Aristotle failed to do. He discovered from the alphabet onward, the great vortices of power at the interface of cultural frontiers. He recovered for the West the world of entelechies and formal causality long buried by the logicians and teachers of applied knowledge; and he did this by looking carefully at the immediate situation created by staples and the action of the Canadian cultural borderline on which he was located.

Looking for the missing face of Canada, Kenner feels Canadians have been beguiled into nonentity by the appeal of the big, tough "wilderness-tamers" and our urge to identify with "rock, rapids, wilderness and virgin (but exploitable) forest." But Kenner gave up just when the trail was promising. Yes, the Canadian, as North American, answers the call of the wild and goes out into the wilderness to "invite his soul," but unlike the rest of mankind, he goes out with a merely private face (and also a private voice). Whereas the Frenchman or the Russian or the Irishman records the defeats and miseries (as well as the joys and successes) of his life on his countenance, the North American keeps his face to himself and "scrubs" it daily.

Somewhat in the manner of Dorian Gray, the real picture of the individual life is hidden away for private judgment rather than public inspection. On the other hand, the extrovert who goes outside to be a lonely fighter and explorer is not an extrovert at home. Charlie Chaplin's pictures of the lonely tramp never take us inside an American home, thereby ignoring the hidden ground of his lonely figure of the Little Tramp. Chaplin was an Englishman who never understood America, but he gave Europeans what they still view as American documentaries.

Equally as fascinated and confused as Chaplin, W. H. Auden shared the bafflement of Henry James about the missing face in North America:

> *So much countenance and so little face.* (Henry James) Every European visitor to the United States is struck by the comparative rarity of what he would call a face, by the frequency of men and women who look like elderly babies. If he stays in the States for any length of time, he will learn that this cannot be put down to a lack of sensibility—the American feels the joys and sufferings of human life as keenly as anybody else. The only plausible explanation I can find lies in his different attitude to the past. To have a face, in the European sense of the word, it would seem that one must not only enjoy and suffer but also desire to preserve the memory of even the most humiliating and unpleasant experiences of the past.
>
> More than any other people, perhaps, the Americans obey the scriptural injunction: "Let the dead bury their dead."
>
> When I consider others I can easily believe that their bodies express their personalities and that the two are inseparable. But it is impossible for me not to feel that my body is other than I, that I inhabit it like a house, and that my face is a mask which, with or without my consent, conceals my real nature from others.
>
> It is impossible consciously to approach a mirror without composing or "making" a special face, and if we catch sight of our reflection unawares we rarely recognize ourselves. I cannot read my face in the mirror because I am already obvious to myself.
>
> The image of myself which I try to create in my own mind in order that I may love myself is very different from the image which I try to create in the minds of others in order that they may love me.[9]

Auden is here speaking of a psychic dichotomy alien to North Americans. "The case of the missing face," however, has a simple solution from one point of view, since the North American, in poetry, art, and life, tends to substitute the face of nature for the human countenance.

Like Wordsworth and Thoreau North Americans spend their time scanning the environmental mystery, taking spins in the country instead of spinning thoughts at home. The North American goes to the movies or theater to be alone with his date, whereas Europeans go to enjoy the audience. The North American excludes advertisements from his cinema and theater, while Europeans find no violation of their privacy from ads in places of public entertainment. Europeans, on the other hand, exclude ads from radio and television in their homes; but since there is little or no privacy in the North American home, ads are tolerated, if only because we go elsewhere for privacy.

Earlier I noted that the entire paradox of the "reversed space" of the North American frustrated Henry James who made it a psychological crux in his novels, regarding it as "the complex fate" of being American. In *Daisy Miller,* a story of an American girl in Europe, he noted:

> She has that charming look that they all have, his aunt resumed. I can't think where they pick it up; and she dresses in perfection—no, you don't know how well she dresses. I can't think where they get their taste.
>
> But, my dear aunt, she is not, after all, a Comanche savage.
>
> She is a young lady, said Mrs. Costello, who has an intimacy with her mamma's courier.

She was not accepted into the space of European society and is considered a kind of "noble savage":

> Winterbourne stood looking after them; he was indeed puzzled. He lingered beside the lake for a quarter of an hour, turning over the mystery of the young girl's sudden familiarities and caprices. But the very definite conclusion he came to was that he should enjoy deucedly "going off" with her somewhere.

They do indeed go off to the Castle of Chillon, and this also increases his puzzlement:

> The sail was not long, but Winterbourne's companion found time to say a great many things. To the young man himself their little excursion was so much of an escapade—an adventure—that, even allowing for her habitual sense of freedom, he had some expectation of seeing her regard it in the same way. But it must be confessed that, in this particular, he was disappointed.

Going out with her date to be alone was in fact the most natural thing in the world for Daisy Miller, whereas for her European friend, Winterbourne, it was a bizarre event:

> Daisy Miller was extremely animated, she was in charming spirits; but she was apparently not at all excited; she was not fluttered; she avoided neither his eyes nor those of any one else; she blushed neither when she looked at him nor when she felt that people were looking at her. People continued to look at her a great deal, and Winterbourne took much satisfaction in his pretty companion's distinguished air. He had been a little afraid that she would talk loud, laugh overmuch, and even, perhaps desire to move about the boat a good deal. But he quite forgot his fears; he sat smiling, with his eyes upon her face, while, without moving from her place, she delivered herself of a great number of original reflections. It was the most charming garrulity he had ever heard. He had assented to the idea that she was "common;" but was she so, after all, or was he simply getting used to her commonness?[10]

Henry James's delight in presenting his American personae to European bewilderment is a very rich subject indeed, but one which depends entirely on understanding the peculiar American quest for solitude in the wilderness, and discovering thereby a privacy and a psychic dimension which Europeans cannot encompass or understand. The conflict in the mind of Henry James had occurred earlier in the life and work of Hawthorne. The latter had also confused the North American quest for privacy out-of-doors with a weak concession to autocratic values and thereby a betrayal of democratic values. In going outside to be social, the European seemed more democratic than the North American going outside to be alone:

> To state the case succinctly: Hawthorne's compulsive affirmation of American positives, particularly in the political sense, led to a rejection of the idea of solitude; and solitude as an expression of aristocratic withdrawal sided with Europe rather than America when the two traditions stated their respective claims.[11]

Henry James finally clarified the conflict by a confession about his personal life which he confided to Hamlin Garland:

> He became very much in earnest at last and said something which surprised and gratified me. It was an admission I had not expected

him to make. "If I were to live my life over again," he said in a low voice, and fixing upon me a somber glance, "I would be an American. I would steep myself in America, I would know no other land. I would study its beautiful side. The mixture of Europe and America which you see in me has proved disastrous. It has made of me a man who is neither American nor European. I have lost touch with my own people, and I live here alone. My neighbors are friendly, but they are not of my blood, except remotely. As a man grows old he feels these conditions more than when he is young. I shall never return to the United States, but I wish I could."[12]

Refusing to accept the European past he had not earned or made, Henry James is here telling Canadians that the American sense of identity was as much a question mark in the late nineteenth century to Americans as the Canadian sense of identity is to Canadians today, and James serves to stress the crucial role of the imaginative native artist in creating the uncreated consciousness of a people.

It is by an encounter with the hidden contours of one's own psyches and society that group identity gradually develops. That Canada has had no great blood-letting such as the American Civil War, may have retarded the growth of a strong national identity, reminding Canadians that only the bloody-minded could seriously wish to obtain a group identity by such violence.

The 1976 strike of Canadian air pilots and controllers over the bilingual issue at airports clearly marks another of the vivid borderlines of Canadian interface and abrasion. The French language is a cultural border and vortex of energy that has roots in the beginnings of Canada as a territory cherished by both French and English settlers. The new technology of air travel projects an ancient quarrel in a new dramatic medium. The drama of the civilian air controllers resonates with the events of 1759 and the fall of Quebec City. The repercussions of that event affect all of North America today. Donald Creighton discerns the action on both sides of the border in the opening sentence of *The Empire of the St. Lawrence:*

When, in the course of a September day in 1759, the British made themselves the real masters of the rock of Quebec, an event of apparently unique importance occurred in the history of Canada.

There followed rapidly the collapse of French power in North America and the transference of the sovereignty of Canada to Great Britain; and these acts in the history of the northern half of the continent may well appear decisive and definitive above all others. In fact, for France and England, the crisis of 1759 and 1760 was a climax of conclusive finality. But colonial America, as well as imperial Europe, had been deeply concerned in the long struggle in the new continent; and for colonial America the conquest of New France had another and a more uncertain meaning. For Europe the conquest was the conclusion of a drama; for America it was merely the curtain of an act. On the one hand, it meant the final retirement of France from the politics of northern North America; on the other, it meant the regrouping of Americans and the reorganization of American economies.[13]

The Fall of Quebec (1759) and the Peace of Paris (1763) created the same psychic border for French Canada as the Civil War defeat did in the mind of the American South. The defeat stimulated the feeling of an historical present that was absent in the victors. "For many French Canadians," writes Ramsay Cook in *The Maple Leaf Forever,* "the past, and especially the conquest, has always been part of the present." He continues with the words of Canon Groulx:

History, dare I say it, and with no intention of paradox, is that which is most alive; the past, is that which is most present. Or in Esdras Minville's revealing remark about "we who continue history, who are history itself." This attitude toward history which makes the past part of the present is not, of course, uniquely French Canadian. It bears a marked similarity to the comment of a distinguished Mexican philosopher concerning Hispanic America. "The past, if it is not completely assimilated, always makes itself felt in the present," Leopoldo Zea has written, "Hispanic America continued to be a continent without a history because the past was always present. And if it had a history, it was not a conscious history. Hispanic America refused to consider as part of its history a past which it had not made." Is it not the failure to "assimilate" the Conquest, to make it French-Canadian history, that explains the endless attempts to interpret it?[14]

These hidden borders in men's minds are the great vortices of energy and power that can spiral and erupt anywhere; and it is not for lack of such vortices that the Canadian identity is obscure.

Rather, there are so many that they have been dissipated and smothered in consumerism and affluence. The vast new borders of electronic energy and information created by radio and television have set up world frontiers and interfaces among all countries on a new scale that alter all pre-existing forms of culture and nationalism. The superhuman scale of these electric "software" vortices has created the Third World with its threat to the old industrial world of "hardware."

On the occasion of Queen Elizabeth's visit to the White House to congratulate the United States on its bicentennial, the president proposed that the dignity of the occasion might be enhanced by reading aloud the Declaration of Independence. So much for subliminal wisdom, even though a G. K. Chesterton might well have found in this much food for transcendental meditation. The historian Kenneth McNaught observes:

> It is sometimes said that Americans are benevolently uninformed about Canada while Canadians are malevolently well-informed about the U.S.[15]

McNaught is here pointing to one of the great borderline features of the Canadian, namely, his opportunities to "take over" the United States intellectually in the same way Alexis de Tocqueville did earlier. Is it not significant that Tocqueville was unable to see the French situation with the same clarity that he brought to the United States? France was driving to the extremes of specialism and centralism under the fragmenting pressure of print technology, creating the matrix for Napoleon while Tocqueville was enjoying the naiveté of Americans whose politics were the first to be founded on the printed word.

In the same way, Canadians repine in the shadow of the American quest for identity, saying "Me too!" while ignoring the anguish of the American struggle to find out "who are we?" Kenneth McNaught cites the American historian C. Vann Woodward to underline the plight of fellow borderliners:

> How many of us have experienced a feeling of being really Canadian only during our first trip abroad? I suspect that this feeling flows from the fact that we have always been self-confident without really understanding it.

Many Canadians had their first vivid experience of national identity while watching the Russian teams play the Canadians at their own game of hockey. The Russians gave them a very bad time by playing a close style of hockey that the Canadians had developed in the 1930s and forgotten in the 1940s. It was in the earlier period that the Canadians sent their coaches to Russia to teach the game. Here was surely an admirable example of the borderline case in full interface. Canadian participation in past wars, whether in 1812 or 1914 or after, has never been on a scale to enable them to identify with the total operation. With hockey the scale is right but the personnel is confusing. In the Olympics the American hockey team consists mostly of Canadians. The jet plane knows no geographic borders with the result that hockey is played by Canadians in American arenas as an American sport. There is enough psychic and social overlap to make both American baseball and Canadian hockey acceptable dramatizations of the competitive drives and skills of both countries.

Related and comparable in scope to the gap of the missing face of North America is the case of the missing voice. If going outside to be alone forbids the assuming of a culturally acquired countenance, the same inhibition extends to "putting on" the North American voice. When we go outside, we use only a private voice and avoid the cultivation of an educated or modulated tone. One might even suggest that the absence of class barriers in North America owes more to its refusal to assume a group or class speech than to its political convictions or institutions.

When a Bill Buckley, Jr., tilts his head and intones on television, he is clowning. Any American who tried to do seriously what the British public schoolboy is taught to do publicly would be run out of town on a rail. The North American hesitation to "put on" a public voice or a face is also a block to the artist and writer in "putting on" an audience for his work. Going out to be alone is antithetical to the role of the artist who must invent an image that will sting or intrigue a public to encounter his challenge. For the artist has to upset his audience by making them aware of their automatism or their own inadequacy in their daily lives. Where mere survival exhausts the creative energies as on our borderline, few have the daring to confront their public with an aesthetic vision.

There is a great new TV borderline in North America which endangers many established features of our lives, including our assumed right to use only the private voice out-of-doors. The TV generation has begun to "put on" a peer group or tribal dialect which could send us "Upstairs *or* Downstairs" in our sleep. The fact that the great vortex of interface between inside and outside in North America has gone unnoted by historians and psychologists for two centuries and more, is testimony to the vast subliminal energies that are outside our consciousness.

Canada is a land of multiple borderlines, psychic, social, and geographic. Canada has the longest coastline in the world, a coastline which represents the frontier for Europe on one side and the Orient on the other side. T. S. Eliot was very conscious of the "updating" power of frontiers. Commenting on this in regard to Mark Twain and the Mississippi on whose shores he was born, Eliot states:

> . . . I am very well satisfied with having been born in St. Louis: in fact, I think I was fortunate to have been born here, rather than in Boston, or New York, or London.[16]

Within the city itself, he was conscious of borderlines. Mentioning the boundaries of the city, he says:

> . . . the utmost outskirts of which touched on Forest Park terminus of the Olive Street streetcars, and to me, as a child, the beginning of the Wild West.

Eliot is especially concerned with the effects of borderlines on language and literature, seeing in Mark Twain:

> . . . one of those writers, of whom there are not a great many in any literature, who have discovered a new way of writing, valid not only for themselves but for others. I should place him, in this respect, even with Dryden and Swift, as one of those rare writers who have brought their language up to date, and in so doing, "purified the dialect of the tribe."[17]

Eliot sees the frontier as an area of transformation and purgation, a character which belongs to frontiers and borderlines in many other places. Frederick J. Turner wrote a celebrated piece "The Significance of the Frontier in American History," noting that:

American social development has been continually beginning over again on the frontier. This perennial rebirth, this fluidity of American life, this expansion westward with its new opportunities, its continuous touch with the simplicity of primitive society, furnish the forces dominating American character.[18]

The Canadian borderline (as well as the numerous frontiers within Canadian borders) shares many of the features that Turner observes concerning the frontier in American history. He sees it as reacting on Europe as well as being the door for the entry of Europeans. In saying that the frontier is "the line of most rapid and effective Americanization" he is pointing to one of the major features of the Canadian borderline where the process of Canadianization also takes place. A frontier, or borderline, is the space between two worlds, constituting a kind of double plot or action that the poet W. B. Yeats discovered to be the archetypal formula for producing "the emotion of multitude" or the sense of universality. In his essay "The Emotion of Multitude" Yeats explains and illustrates how in poetry and in art, the alignment of two actions without interconnection performs a kind of magical change in the interacting components. What may be banal and commonplace situations, merely by their confrontation and interface, are changed into something very important.

The borderline of interface between cowboys and Indians captured the imagination not only of Europeans, but of Orientals as well. Even today affluent Germans and Japanese dress up in the costumes of cowboys and Indians and mount their horses to play the games of the Wild West. This frontier, to them exotic, has always been a major part of the Canadian experience. The old frontier had been the melting pot for the immigrant, while today it continues as the melting pot of the affluent. Suburban Canada camps and cottages in the North.

Borderlines, as such, are a form of political "ecumenism," the meeting place of diverse worlds and conditions. One of the most important manifestations of Canadian ecumenism on the Canadian borderline is the interface between the common law tradition (oral) and the American Roman law (written). There have been no studies of this very rich situation, but then there have been no studies of how the oral traditions of the Southern states are the creative foundations of American jazz and rock music. Borderlines maintain an attitude of alertness and mutual study which gives a cosmopolitan character to Canada.

One of the more picturesque borderlines in Canadian life is its royal commissions which serve as mobile interdisciplinary and intercultural seminars, constituting a kind of "grass roots" tradition in Canada. In search of hyperbole, only a Canadian can say: "As Canadian as a Royal Commission!" There is an outré hyperbole of even more local significance: "As Canadian as Diefenbaker's French!" John Diefenbaker would have been delighted to know that he had become a cultural frontier!

Yes, Canada is a land of multiple borderlines, of which Canadians have probed very few. These multiple borderlines constitute a low-profile identity, since, like the territory, they have to cover a lot of ground. The positive advantage of a low profile in the electronic age would be difficult to exaggerate. Electronic information now encompasses the entire planet, forming another hidden borderline or frontier whose action has been to rob many countries of their former identities. In the case of the First World, the Fourth World of electronic information dims down nationalism and private identities, whereas in its encounter with the Third World of India, China, and Africa, the new electric information environment has the effect of depriving these people of their group identities. The borderline is an area of spiraling repetition and replay, both of inputs and feedback, both of interlace and interface, an area of "double ends joined," of rebirth and metamorphosis.

Canada's 5000-mile borderline is unfortified and has the effect of keeping Canadians in a perpetual philosophic mood which nourishes flexibility in the absence of strong commitments or definite goals. By contrast, the United States, with heavy commitments and sharply defined objectives, is not in a good position to be philosophic or cool or flexible. Canada's borderline encourages the expenditure on communication of what might otherwise be spent on armament and fortification. The Canadian Broadcasting Corporation and the National Film Board are examples of federally sponsored communication rather than fortification. At the same time, Canadians have instant access to all American radio and television which, experienced in the alien milieu of Canada, feeds the philosophic attitude of comparison and contrast and critical judgment. The majority of Canadians are very grateful for the free use of American news and entertainment on the air, and for the princely hospitality and neighborly dialogue on the ground.

The advantages of having no sharply defined national or private identity in Canada appear in the general situation where lands long blessed by strong identities are now bewildered by the growing preformation and porousness of their identity image in this electronic age. The low-profile Canadian, having learned to live without such strongly marked characteristics, begins to experience a security and self-confidence that are absent from the big-power situation. In the electronic age centralism becomes impossible when all services are available everywhere. Canada has never been able to centralize because of its size and small population.

The national unity which Canadians sought by the railway "hardware" now proves to be irrelevant under electronic conditions which yet create an inclusive consciousness. For Canada a federal or inclusive consciousness is an inevitable condition of size and speed of intercommunication. This inclusiveness, however, is not the same as the nineteenth-century idea of national unity; rather, it is that state of political ecumenism that has already been mentioned as the result of multiple borderlines.

In order to have a high-profile identity nationally and politically, it is necessary to have sharp and few political and cultural borders. From 1870 onward Germany strove for a high-profile identity within its multiple borders. In the industrial age this drive toward centralized and intense identity imagery seemed to be part of competitive commerce. Today, when the old industrial hardware is obsolescent, we can see that the Canadian condition of low-profile identity and multiple borders approaches the ideal pattern of electronic living.

A GLOSSARY OF TETRADS
Tracing the Shift from Angelism (Visual Space) to Robotism (Acoustical Space)

Tetradic Glossary

Equilibrium

(A) Uncertainty: any input amplifies or inflates some situation
(B) Fixity: obsolesces existing homeostasis or balance
(C) Ground: re-creates an older mode of equilibrium
(D) Progressive motion: when pushed to its limits, the system reverses it modalities

Visual Space

(A) Amplifies continuum: space as container (Euclid-Newton)
(B) Obsolesces connectedness
(C) Brings back homogeneity
(D) Reverses into steady state condition: acoustic space

Perspective

(A) Enhances private point of view
(B) Obsolesces panoramic scanning
(C) Retrieves specialism
(D) Reverses into cubism, multi-view

Number

(A) Amplifies plurality; quantity, for example, possessions
(B) Obsolesces notches, ideographs, tallies
(C) Creates math operations: zero, blank, algebraic singularity
(D) Reverses into profile of crowd: pattern recognition

Clock

(A)   Amplifies work via the storing of mechanical energy
(B)   Obsolesces leisure in the time-regulated city
(C)   Retrieves history as an art form; human memory set down
      through fixed chronology (battles, defeats—1066, 1763,
      1945, etc.)
(D)   Reverses into the eternal present via simultaneous pattern rec-
      ognition (myth), such as the seventeenth-century "Sacra-
      ment of the Present Moment" Benedictine monks: "Labo-
      rare Est Orare"

Copernican Revolution

(A)   Enhances role of the sun (central)
(B)   Pushes aside the crystalline spheres
(C)   Retrieves theories of Aristarchus of Samos (275 B.C.)
(D)   Flips into relativity—centers everywhere and margins nowhere
      (acoustic space)

Periodic Tables

(A)   Intensifies classification
(B)   Pushes out alchemy
(C)   Retrieves the idea of families and of structures; reopens search
      for underlying unity
(D)   Reverses into wave theory of Erwin Schrodinger

Atomic Structure

(A)   Enhancement by atomic combination
(B)   Obsolesces the four elements of the Greeks
(C)   Retrieves the ancient atomist theory
(D)   Flips into contemporary atom of Leucippus and Democritus in
      the form of solar structure (Newtonian theory/Niels Bohr)

Mirror (*Mirari*: to wonder)

(A)   Enhances ego by repetition and self-advertisement; echo-match-
      ing of a figure-minus-its-ground instrument for self-
      portraiture—Rembrandt, etc.; adjunct of phonetic literacy
      via visual intensity: tunnel vision

(B)   Obsolesces the corporate mask and corporate appearance (costume); personal dress replaces costume
(C)   Retrieves the mode of Narcissus (magic, metaphoric tunnel vision) (self-portrait: mirror as sitter, painter as audience and as admirer)
(D)   Reverses into "making" process as recognition, replay; outlook becomes insight

Metaphor

(A)   Enhances awareness of relations
(B)   Obsolesces simile, metonymy, connected logic
(C)   Retrieves understanding, "meaning," via replay in another mode
(D)   Reverses into parallelism, allegory

Spoken Word (Mirror of the mind: *canon* is mirror of the voice, when one voice repeats or reflects

(A)   Enhances self-awareness: consciousness
                          what another has stated)
(B)   Obsolesces the sub-human
(C)   Retrieves past experience
(D)   Group competitiveness and class structure

Printed Word

(A)   Amplifies private authorship, the competitive goal-oriented individual
(B)   Obsolesces slang, dialects, and group identity, separates composition and performance, divorce of eye and ear
(C)   Retrieves tribal elitism, charmed circle, cf., the "neck verse"
(D)   With flip from manuscript into mass production via print comes the corporate reading public and the "historical sense"

Crowd

(A)   Intensifies need to increase
(B)   Obsolesces private identity
(C)   Retrieves paranoia
(D)   Reverses into violence: fear of decrease

Clothing

(A) Amplifies private energy: clothing as weaponry
(B) Obsolesces climate: clothing as thermal control
(C) Brings back mask, trophy, group (corporate) energy
(D) Flips into conventional attire (IBM dress code)

Housing

(A) Private enclosed visual space ("Three Little Pigs")
(B) Cave, tent, wigwam, dome
(C) Wagon trains, covered wagons (pioneers), mobile home clusters
(D) Corporate identity in high-rise

City

(A) Intensifies the centralizing of all human activities
(B) Obsolesces the countryside, the rural
(C) Retrieves homeostasis—"bustle"
(D) Reverses into suburb: breakdown of centralism

Elevator

(A) For mines: enhanced depth—real "low down"
(B) Steps, ladders—gravity, that is, levity
(C) Retrieves hidden treasures as well as hierarchy
(D) Flips into high-rise: new egalitarianism of elevator

High-Rise (Skyscraper)

(A) Amplifies privacy
(B) Obsolesces community
(C) Retrieves catacomb—an apartment is not a home
(D) Reverses into slum: community in crisis

The Wheel

(A) Accentuates locomotion
(B) Obsolesces sled, roller, greased skids, etc.
(C) Retrieves roads as rivers (moving sidewalk), skis, snowmobiles
(D) Reverses into airplane, via bicycle

Compass

(A)   Enhances range and accuracy of navigation
(B)   Obsolesces stars
(C)   Retrieves astronomy as art form
(D)   Reverses into electric environment: circuitry as exo-nervous system (cosmic ground)

Cash Money

(A)   Speeds transactions
(B)   Obsolesces barter
(C)   Retrieves conspicuous consumption
(D)   Reverses into credit or non-money

Credit

(A)   Enhances inflation, through indebtedness
(B)   Obsolesces sole ownership; encourages rent-all
(C)   Retrieves cashless society; brings back barter and do-it-yourself
(D)   Flips into bankruptcy

Pension

(A)   Enhances image of future security
(B)   Obsolesces thrift as survival mechanism
(C)   Retrieves "Garden of Eating" (consumerism)
(D)   Pushed far enough results in indigence

Gun Powder

(A)   Extends the range of any steel-cased projectile
(B)   Obsolesces individual personal combat
(C)   Brings back the "Superman"—the group charge
(D)   Flips into total automated death

Steamboat

(A)   Opened the sea for hardware
(B)   Obsolesced the wood/sail craft, fostered uncertainty and exploration
(C)   Created tourism: programmed pilgrims

(D)   Flipped to centralism via iron sea power (vs. old decentraliza-
       tion of sail sea power)

Railway

(A)   Improves horizontal locomotion; increases speed
(B)   Obsolesces the sled, roller, wagon, stage
(C)   Brings frontiers within reach; retrieves ease of river traffic, like
       moving sidewalk
(D)   Reverses into the airplane, via bicycle

Telegraph

(A)   Amplifies the isolated incident into an inclusive dateline; shifts
       front-page content from analysis to instant reporting
(B)   Obsolesced the Addison and Steele style newsheet, and the pri-
       vate point of view
(C)   Retrieved corporate or group involvement, whereby, for exam-
       ple, Baltimore was instantly apprised of Washington con-
       gressional events (1844)
(D)   Reverses into a dynamic broadcasting mode; newspaper front
       page becomes a mosaic of unrelated time-based items

Camera

(A)   Snapshot enhances aggression of individual user; Locke model
       of mind as reflection
(B)   Obsolesces privacy of subjects, providing ego trip
(C)   Retrieves past as present; retrieves tribal corporate image state;
       ego trip for subject
(D)   Reverses into public domain—photo journalism and cinema

Electric Light

(A)   Amplifies space as a visual figure and turns it into ground; in-
       stant night into day
(B)   Obsolesces the mystery of the nonvisual; also, candles, lamps,
       oil, gas
(C)   Retrieves daytime activities on grand scale, i.e., night baseball;
       puts outer (sun) light inside for detailed manipulations,
       e.g., brain surgery
(D)   As Lusseyran says, reversal is blinding: outer vision is con-
       verted to inner trip; figure and ground merge

Automobile

(A) Enhances privacy: people go out in their cars to be alone
(B) Obsolesces the horse-and-buggy, the wagon
(C) Retrieves a sense of quest: knight in shining armor
(D) Pushed to its limit, the car reverses the city (urb) into the ex-urb (suburbs); brings back walking as an art form

Zipper

(A) Amplifies grip, clasp
(B) Obsolesces buttons, snaps
(C) Brings back classic, flowing robes
(D) Flips into (velcro) adhesive

Airplane

(A) Amplifies vertical and horizontal locomotion
(B) Obsolesces the wheel and the road, the railway and the ship
(C) Brings back aerial perspective with the aura of miniaturization
(D) Reverses into guided projectile; transforms planet into extended city; urb orbs

Electric Media

(A) Amplification of scope of simultaneity and service environment as information
(B) Obsolesces the segmented visual, connected, and logical
(C) Retrieves the subliminal, audile-tactile dialogue
(D) Etherealization: the sender gets sent

Microphone—PA System

(A) Amplifies individual speech and tonal variety
(B) Obsolesces the big band, the Latin mass, grand opera
(C) Brings back group participation; rhythmic replay
(D) Flips from private to corporate sound-bubble

Radio-Television

(A) Improves (regional) simultaneous access to entire planet— everybody: "On the air you're everywhere"

(B)   Obsolesces wires, cables, and physical bodies
(C)   Retrieves tribal ecological environments: echo, trauma, paranoia, and also brings back primacy of the spatial, musical, and acoustic
(D)   Reverses into global village theater (Orson Welles's *Invasion From Mars:* no spectators, only actors)

Xerox

(A)   Increases speed of printing process
(B)   Obsolesces assembly-line book
(C)   Brings back oral tradition, the committee
(D)   Reversal is "everybody a publisher"

Instant Replay

(A)   Instant replay of experience equals the cliché; amplifies cognitive awareness
(B)   Wipes out the merely representational and chronological
(C)   Retrieves "meaning" (I. A. Richards)
(D)   Flips from individual experience to pattern recognition, the nature of the archetype

Committee

(A)   Enhances group image of authority
(B)   Obsolesces individual responsibility
(C)   Brings back dialogue
(D)   Reverses from specialized job to corporate role

Telephone

(A)   Enlarges the impact and speed of the private voice
(B)   Erodes the body as hardware; creates the disembodied consciousness
(C)   Retrieves sense of telepathy
(D)   Reverses into the party-line; omnipresence, like the conference call or teleconferencing

Computer

(A)   Accelerates logical sequential calculations to speed of light
(B)   Erodes or bypasses mechanical processes and human logic in all sequential operations

(C) Highlights "numbers is all" philosophy, and reduces numbering to body count by touch

(D) Flips into the simultaneous from the sequential; accentuates acoustic over visual space to produce pattern recognition

Cable TV

(A) Amplifies quality and diversity of signal pickup

(B) Obsolesces diffusion broadcasting

(C) Retrieves early transmission broadcast pattern point-to-point (ship to shore)

(D) Reversal is flip to home broadcasting

Teletext

(A) Printed radio: enhances headline service, like early radio, e.g., H. V. Kaltenborn

(B) Obsolesces prolonged TV watching; viewer uses service selectively for short intervals

(C) Retrieves silent film dialogue card, pictograph or pun style; teletype format

(D) Reverses into interactive video-text; qualitative data filter, via data bank

Electronic Fund Transfer (Data base)

(A) Enhances flow of goods and services via phone lines and data base

(B) Obsolesces barter and cash money

(C) Retrieves ostentatious show, along with credit overload

(D) Reverses into an intense state of creditworthiness as pure status (non-money)

Satellite

(A) Enlarges global information exchange

(B) Obsolesces language for images (digital over analog)

(C) Retrieves world view, like earlier compass

(D) Reverses into iconic fantasies

Global Media Networking

(A)   Instantaneous diverse media transmission on global basis: simultaneous planetary feed and counter-feed

(B)   Erodes human ability to code and decode in real time

(C)   Brings back Tower of Babel: group voice in the ether

(D)   Reverses into loss of specialism; programmed earth

Notes and References

For brevity, the main references herein are referred to under the author's name and main title, except where further particulars are needed for identification. Full details will be found in the bibliography.

Chapter 1. The Resonating Interval

1. Left and right brain cognitively function diachronically and synchronically. Though the hemispheres are asymmetrical in sensory preference they cooperate for psychic unity. "Diachronic" is here meant to denote the experiencing of an idea or object in sequential time (i.e., day by day). "Synchronic" means the collective experiencing of an idea and object over a period of years (i.e., from epoch to epoch). Such is the difference implied in analytical psychology between individual consciousness and Jung's "collective unconscious." Cf. Joseph Bogen's ruminations concerning the opinions of such diverse figures as Jerome Bruner, Joan Miro, and Henry Moore on the role of the appositional synchronic mind in creativity; "The Other Side of the Brain, III: The Corpus Callosum and Creativity," pp. 198–202. Also consult Barrington Nevitt's review of synchronic and diachronic structures in the work of Descartes, Saussure, and Lévi-Strauss, *The Communication Ecology*, pp. 51–57.

2. Resonating interval as borderline: The bias of visual space—diachronic—is related to the sensory preference of labeling and hierarchy in the left hemisphere (the linear-quantitative). The bias of the right hemisphere is primarily gestalt—synchronic—or pattern formation (configurational-qualitative), i.e., the singular element as opposed to the holistic. The propositional/diachronic mind cooperating with the appositional/synchronic mind enables the human consciousness to di-

vine past, present, and future as inherent in any artifact as T. S. Eliot intimates in "Burnt Norton," The Four Quartets (1934):

> Time present and time past
> Are both perhaps present in time future
> And time future contained in time past.
> If all time is eternally present
> All time is unredeemable . . .

Source: Conversation with Marshall McLuhan, April 21, 1978.

3. In so far as the tetrad reveals the whole life of the artifact simultaneously, it should be considered a right-brain phenomenon (configurational-qualitative). The fully matured tetrad always reveals four process pattern states in apposition (A : B as C : D), but acting in a general position of complementarity not polarity. The abrading relationship, for example, of enhancement and obsolescence, as defined in the McLuhan hypothesis, has a correlative in the poet's efforts in relation to tradition, e.g., every truly new poem recalls the past and alters the historical context of all poetry from Homer to the present day. The use of an artifact today inevitably retrieves its uses in the past in a new context: nostalgia. Cf. T. S. Eilot's essay "Tradition and the Individual Talent," in *Selected Essays* (1950), pp. 3–11, esp. p. 4. Each mature tetrad contains two figure-ground situations (enhancement and obsolescence; retrieval and reversal) in roughly complementary balance, which if seen simultaneously creates an insight called by McLuhan, "comprehensive awareness." Comprehensive or integral awareness is a capacity, not found where the bias of visual space is dominant, to perceive all figure-ground representations simultaneously; similar to the perception of ancient Chinese wisdom expressed in the "cyclic nature of . . . ceaseless motion and change . . ." in the yin/yang principle of the Tao. Cf. Capra, *The Tao of Physics,* pp. 106–108. Also refer to Powell's *The Tao of Symbols,* pp. 102–110; *Oracle Bones, Stars and Wheelbarrows,* by Frank Ross, Jr., pp. 28–35.

4. In the McLuhan hypothesis, following Rubin, the term *figure* is consistently a term of left-brain cognition, in the sense of J. E. Bogen's "two minds in one skull." To isolate the figure as an area of psychic attention is usually to separate out single elements from the total environment. McLuhan usually used the term *ground* to mean a peculiarity of right-brain cognition which senses all figures in the entire environmental surround at once (configuration: a massing of figures). In tetradic analysis every new artifact, whether idea or object, reshapes the environment as it impacts upon it as figure against ground; yet, at the same time, the ground is being altered and eventually reshapes how the artifact is used. See James Striegel's unpublished doctoral

thesis, "Marshall McLuhan on Media," ch. IV, "The Analogical Model," esp. pp. 100–114.

5. On synesthesia see Gombrich, *Art and Illusion*, pp. 366–368, also pp. 370–371, 373–375; on learning and art as representation, not replica, cf. pp. 172–174, 308, 320–321, 324, 356–358, 394.

6. Formal cause vs. efficient cause: In the early Greek courts of law, the concept of formal cause was that it was a defining formula or definition of a thing's essence (its form or "whatness," whereby we know a thing). Somewhere in the centuries before the Renaissance, Aristotelian formal cause was abandoned for efficient cause. Formal cause concerns itself with the continual transformation of all elements in our environment, our habitat, in human terms. The question of formal cause among the Greek elders almost always centered around the idea of whether the artifact was worthwhile in moral and ethical terms. Efficient cause, then as now, simply describes the artifact as to its technical use, without respect to human consequences (i.e., the rubric of scientific determinism). Marshall McLuhan stated that since all artifacts are an extension of the user (cf. *Understanding Media*) then all artifacts are a form of human speech, of language. The ancient sense of Logos, even before Aristotle, postulated that at the moment of utterance the artifact was created in the mind, even though its physical form had not yet appeared—an event dictated by the evolutionary state of the culture, as all artifacts define themselves in relation to ground. The Chinese, for example, initially used gunpowder for religious ceremonies, not war.

7. When man left biology for technical evolution (bows and arrows are as much a part of evolution as losing our tails), artifacts began to function like words to bridge the gap between the replacement of one technology for another (Greek *metapherein* and Latin *transferre:* to carry across). Artifacts, as metaphors, express the dominant concepts of the age, summarizing and gathering together those key metaphors characterizing the consciousness of a particular culture. Barrington Nevitt in *The Communication Ecology*, p. 104, organizes McLuhan's thought on the technological metaphor as follows:

> At every stage of human history, the dominant technologies reverberate in current metaphors which translate unfamiliar aspects of existence into familiar forms. . . . In biblical times the allusions were to agriculture and household arts, to fishing and seafaring, and to tribal warfare. Its millennial metaphor was the Garden. After Gutenberg and the Renaissance, there were perspectives and points of view, telescopes and microscopes, waterpower and clockwork, navigation and gunpowder that ushered in the age of reason with its mechanical forces. In the First Industrial Revolution

there were steam-engines, lineal rails, production lines, gradual progress, evolution and missing links. Its centennial metaphor was the Machine. Today in the midst of the Second Industrial Revolution we speak of fields, feedback, quantum leaps, and information traveling at the speed of light.

McLuhan emulated Sapir and Whorf by presuming that language is culture and postulating that artifacts function as words. *"Logos* encompasses both idea and object."

8. New evidence has come to light that memory, as well as other aspects of the mind, are stored over a large portion of the nervous system; and that memory, for example, is "more like a hologram than a photograph. . . ." Cf. Cooper, Leon and Imbert, Michael, "Seat of Memory," *The Sciences* (Feb. 1981).

9. The topology of the moebius strip embodies a new branch of mathematics useful for solving nonlinear differential equations relating to rates of change. Marshall McLuhan thought that such a typology, as a new intellectual emblem, might be useful in expressing the persistently altering relationship of the tetrad's process patterns. The one-sided character of the moebius band could show the stress relationships between the two figure-grounds "double ends joined," as well as the properties of formal cause: synchronic, many-centered and holistic. For the way in which effects may be expressed mathematically in a nonlinear context, consult Albert W. Tucker and Herbert S. Bailey, Jr., "Topology," in *Scientific American* (Jan. 1950), pp. 18–24, esp. p. 24.

10. The double ends joined or and-both percept expressed in the tetrad has long been used by artists as a mirror image. Tetradically, the mirror down through history (*mirari;* to wonder) amplifies the ego by repetition and self-advertisement, as figure minus its ground. The mirror promotes visual intensity, like tunnel vision. The mirror obsolesces a sense of tribal self, the corporate mask, and prompts private, idiosyncratic dress. The more the mirror is used, the more the self is hypnotized into isolation, engendering a revelation of one's hidden feelings. Outlook becomes insight. Vide *Girl Before a Mirror,* Pablo Picasso, 1932; *Portrait of Gala,* Salvador Dali, 1935 (both works are in the Museum of Modern Art, New York).

Chapter 2. The Wheel and the Axle

1. T. S. Eliot, *Selected Essays,* p. 5.
2. McLuhan and Watson, *From Cliché to Archetype,* pp. 118–119.
3. On archetypal unconscious cf. *From Cliché to Archetype,* pp. 21–23.
4. C. G. Jung, *Psyche and Symbol,* p. XVI.

5. Jean Piaget, *Structuralism,* pp. 52–57, esp. p. 57.

6. Media-related cultural shifts in time and space: Harold Innis in *The Bias of Communication* and in *Empire and Communications* made many historical observations on the differing patterns and structures in human organization as they related to different means available for shaping cultural situations. One of his most frequent illustrations of this principle concerned the two types of bureaucracy that grew from the use of stone, on the one hand, and paper, on the other, as materials for writing. When stone, brick, or clay are used as writing materials, the bureaucracy or human organization of interests and energies tends to take on a priestly form dedicated to stability in time. When paper is broadly available, the bureaucracy tends to become military with a strong interest in the control of space. (In *The Technology of Man* Derek Birdsall and Carlo W. Cipolla pointed out that the earliest clay writing, for example, was used by the priestly caste in Sumeria to systematize temple tithing over the years; the early Egyptians eventually used papyrus for voluminous military and diplomatic communications in both cuniform and demotic script.) Innis was not only concerned with the study of changes in the outer patterns of human organization resulting from different means of communication in time and space, but he was much interested in the changes that took place in the perceptual lives of the people involved in those changes. He played the inner and outer aspects of innovation and change back and forth against each other as a figure-ground interface.

7. Tetradic reversal (data overload equals pattern recognition): That is, centralism reverses into decentralism, hardware to software, jobs to role. See Marshall McLuhan and Barrington Nevitt, *Take Today: The Executive as Dropout* (New York: Harcourt, Brace, Jovanovich, 1972).

8. H. J. Eysenck, "Personality and the Law of Effect," in *Pleasure, Reward, Preference,* p. 133.

9. Thomas Kuhn, *The Structure of Scientific Revolutions,* p. 24.

10. Saussure, *Course in General Linguistics,* p. 81.

11. E. R. Leach, *Claude Lévi-Strauss,* p. 15.

12. Leach, *Claude Lévi-Strauss,* pp. 7–8.

13. Leach, *Claude Lévi-Strauss,* p. 22.

14. Leach, *Claude Lévi-Strauss,* p. 23.

15. John C. Lilly, *The Centre of the Cyclone,* p. 67.

16. Anais Nin, *D. H. Lawrence,* p. 33.

17. Jacques Ellul, *Propaganda: The Formation of Men's Attitudes,* p. 15.

18. Ellul, *Propaganda: The Formation of Men's Attitudes,* p. 9.

19. Georges Gusdorf, *Speaking (La Parole),* pp. 42–43.

20. Jacques Lusseyran, *And There Was Light,* p. 144.

21. Nan Lin, *The Study of Human Communication,* p. 192.

22. Aristotle, *De Anima* (Book III, ch. viii), 432^b.

23. C. M. Turbayne, *The Myth of Metaphor,* p. 11. A complete bibliography of the literature is presented in *Metaphor: An Annotated Bibliography and History* by Warren A. Shibles (Whitewater, Wisconsin: The Language Press, 1972).

24. *The Rhetoric of Aristotle* (Book III, ch. IV), p. 192.

25. Paul Ricoeur, *The Rule of Metaphor,* p. 19.

26. Ricoeur, *Rule of Metaphor,* p. 21.

27. Ricoeur, *Rule of Metaphor,* p. 34.

28. On the *enkyklios paideia* cf. H. I. Marrou, *A History of Education in Antiquity,* pp. 176, 210–211. See also, H. I. Marrou, *Saint Augustin et la fin de la culture antique,* pp. 123–124.

29. On the ascendancy of visual stress via the phonetic alphabet, consult *The Gutenberg Galaxy* by Marshall McLuhan for a detailed discussion; also cf. *The Coming of the Book* by Febvre and Martin (London: NLB Atlantic Highlands; Humanities Press, 1976) and *Ramus, Method and Decay of Dialogue,* by Walter Ong (Cambridge: Harvard University Press, 1958).

Chapter 3. Visual and Acoustic Space

1. Lusseyran, *And There Was Light,* pp. 23–24, 48–49.

2. F. M. Cornford, "The Invention of Space," *Essays in Honour of Gilbert Murray,* pp. 215–235.

3. Cicero's training, through Plato's disciples, was influenced by an earlier religious usage that *logos* (a primitive utterance of the word) structured the *kosmos* and infused man's being with a wise concept of world order or common sense. *Heraclitus: The Cosmic Fragments,* ed. Geoffrey S. Kirk (London: Cambridge University Press, 1954), pp. 70, 396, 403. Also, Harold Innis in *Empire and Communications,* p. 76, says "The structure of man's speech was an embodiment of the structure of the world." Cicero's rhetorical theory, as an interchange of both thought and feeling (*inventio, dispositio, elocutio, memoria* and *pronuntia*) became the academic anchor for the medieval trivium; for a form of summation consult Marcus Tullius Cicero, *De Oratore,* trs. E. W. Sutton and H. Rackham (Cambridge: Harvard University Press, 1967), pp. 97–109.

4. Cornford, "The Invention of Space," p. 219.

5. S. Giedion, "Space Conception in Prehistoric Art," *Explorations in Communication: An Anthology,* p. 77.

6. The eternal present: Summarized from an extended exchange

between Edmund Carpenter and Marshall McLuhan during a student discussion of Carpenter's first draft essay "Thinking Through Language," at the Centre for Culture and Technology, University of Toronto. Also, cf. Dorothy Lee, "Lineal and Nonlineal Codifications of Reality," *Explorations in Communication: An Anthology,* pp. 136–154.

7. The Chinese sent trading ships to Africa in 1433, sixty-five years before Vasco da Gama made his exploratory voyages, according to Joseph Needham, Director, East Asian History of Science Library, Cambridge University, England.

8. Eric Havelock, *Origins of Western Literacy,* p. 43.

Chapter 4. East Meets West in the Hemispheres

1. Saussure, *Course in General Linguistics,* pp. 11–13, 77–81, 95, 232.

2. The corpus callosum is a tissue bridge (or hard body) between the two apparently independent hemispheres of the neo-cortex and appears to facilitate the constant neural feedback between the hemispheres which defines the nature of "consciousness" as an integrated operating model of the outside world. Richard M. Restak, *The Brain: The Last Frontier,* pp. 187–202. The corpus callosum as a thick band of nerve fibers joins the left and the right brain (associative or neo-cortex) and coordinates audile-spatial construction and nonverbal ideation (right hemisphere) with calculation, speech, writing, and general linguistic abilities (left hemisphere). Hearing, the direction of left and right visual fields, as well as handedness are associated on a crossover basis between the hemispheres. For most people in Western culture, the left-brain is dominant. The right brain is subordinated ("minor") seemingly involved, in the primary sense, with audile-tactile functions and pattern recognition. Integration between the hemispheres seems handicapped or retarded by the dominance of the left brain. (Book schematics broadly based on the work of Wilder Penfield and Robert Sperry, with ancillary reference to Joseph Bogen.) Note Galin, David, "Hemisphere Specialization: Implications for Psychiatry, *"in Biological Foundations of Psychiatry,* edited by R. G. Grenell and S. Gabay, pp. 154–156.

3. Restak, *The Brain,* pp. 167–169.

4. The brain as mosaic: Neurosurgeon Wilder Penfield's galvanic probes of living human brain tissue revealed that, "The subject feels again (*as the electric probe is inserted*) the emotion which the situation originally produced in him, and he is aware of the same interpretations, true or false, which he himself gave to the experience in the

first place. Thus, evoked recollection is not the exact photographic or phonographic reproduction of past scenes or events. It is a reproduction of what the patient saw and heard and felt and understood." (Wilder Penfield, "Memory Mechanisms," *A.M.A. Archives of Neurology and Psychiatry*, vol. 67 (1952): 178–198).

5. Josephine Semmes, "Hemispheric Specialization: A Clue to Possible Mechanism," *Neuropsychologica*, pp. 11–26.

6. Consciousness and the hemispheres: The area of the recticular formation, in the brain stem, is responsible for the control of wakefulness, the pre-condition of consciousness. By locating the functions in the hemispheres, the authors do not wish to convey the thought that consciousness, as a state, does not involve all parts of the brain as well as elements of the spine. See A. R. Luria, *The Working Brain*, tr. Basil Haigh (1973), pp. 43–67. For a description of consciousness in split-brain studies see Ornstein, Robert E., *The Psychology of Consciousness*, pp. 16–39.

7. Joseph E. Bogen and Glenda M. Bogen, "The Other Side of the Brain, III: The Corpus Callosum and Creativity," *Bulletin of the Los Angeles Neurological Societies*, vol. 34, no. 4 (Oct. 1969): 198–202.

8. R. J. Trotter, "The Other Hemisphere," *Science News*, p. 220.

9. Trotter, "The Other Hemisphere," p. 220, 219 (cf. brain function chart).

10. Brain function chart: R. J. Trotter's chart reflects the scientific understanding of the cortical hemispheres, gained mainly in the last twenty years. The cortex of the ordinary human brain has two hemispheres, joined by a massive bundle of fibers called the corpus callosum, which seems to be the agency of dialogue between the hemispheres. It was only in the 1950s that these forebrain commissures in man were deliberately severed, allowing the hemispheres to be studied independently. Michael S. Gazzaniga, "Review of the Split Brain," in *The Human Brain*, ed. M. C. Wittrock, p. 91, notes:

> The first important finding was that the inter-hemispheric exchange of information was totally disrupted following commissurotomy. The effect was such that visual, tactual, proprioceptive, auditory and olfactory information presented to one hemisphere could be processed and dealt with in that half-brain, but each of these activities went on *outside* the realm of awareness of the other half-cerebrum. This observation confirmed the animal work done earlier by Myers and Sperry, except that in a sense the results were more dramatic. Since it is the left hemisphere that normally possesses the natural language and speech mechanisms, all processes ongoing in the left hemisphere could easily be verbally described by the patients; information presented to the right hemisphere went undescribed.

In subsequent tests of patients that had undergone commissurotomy, the complementarity of the two hemispheres became increasingly evident. Most surprising was the range of activities proper to the right hemisphere, which in the nineteenth century carried the label of "minor" or "quiet." So complete was our culture's visual bias at that time, it was seriously proposed that the right hemisphere made no contribution to human intellection or activity.

11. Trotter, "The Other Hemisphere," p. 218.

12. Acoustic space and the hemispheres: Acoustic space is multisensory, involving the interval of tactility and kinetic equilibrium-pressure, and as such it is one of the many figure-ground right-hemisphere forms of space. Ordinarily the two hemispheres are in constant dialogue through the corpus callosum, and each hemisphere uses the other as its ground except when one (i.e., the left) is habitually dominant. Each hemisphere, as it were, provides a particular type of information processing less available to the other. As one of the surgeons, J. E. Bogen notes "the type of cognition proper to the right hemisphere has been called *appositional,* a usage parallel to the common use by neurologists of *propositional* to encompass the left hemisphere's dominance for speaking, writing, calculation and related tasks." (Joseph E. Bogen, "Some Educational Implications of Hemispheric Specialization," in *The Human Brain,* p. 138). Dr. Bogen observes further: "What distinguished hemispheric specialization is not so much certain kinds of material (e.g., words for the left, faces for the right) but the *way* in which the material is processed. In other words, hemispheric differences are more usefully considered in terms of *process specificity* rather than material specificity."

13. On gestalten cf. Wolfgang Kohler's speculations, as discussed by Jean Piaget in *Structuralism,* pp. 52–57. Also see Wolfgang Kohler, *Gestalt Psychology* (1947), pp. 173–205, *passim.*

14. On the concept of resonance vide Linus Pauling, *The Nature of the Chemical Bond,* p. 12. "The concept of resonance was introduced into quantum mechanics by Heisenberg in connection with the discussion of the quantum states of the helium atom. He pointed out that a quantum-mechanical treatment somewhat analogous to the classical treatment of a system resonating coupled harmonic oscillators can be applied to many systems." Following Heisenberg, Pauling saw acoustic and mimetic resonance as the essential structure of matter. Such resonance, as a guiding principle, should also apply to brain structure neurologically.

Chapter 5. Plato and Angelism

1. Eric Havelock, *Preface to Plato,* pp. 45, 47.
2. Alexander R. Luria, "The Functional Organization of the Brain," in *Scientific American* (March, 1970), pp. 21–71. Harold A. Innis remarked on the Oriental (right-hemisphere) antipathy to sequence and abstraction and Western precision. Cf. *The Bias of Communication,* p. 62.
3. Havelock, *Preface to Plato,* pp. 285–286.
4. Ellul, *Propaganda,* pp. 108–109.
5. Karl Popper, *The Open Society and Its Enemies,* p. 183.
6. Popper, *The Open Society and Its Enemies,* pp. 183–184.
7. Popper, *The Open Society and Its Enemies,* p. 178.
8. Chiang Yee, *The Chinese Eye: An Interpretation of Chinese Painting,* pp. 189–190.
9. Tony Schwartz, *The Responsive Chord,* pp. 14, 16.
10. Herbert Krugman, from a paper delivered to the annual conference of the Advertising Research Foundation, October 1978. See also, Barry Siegal, "Stay Tuned for How TV Scrambles Your Brain," in the *Miami Herald,* Sunday, June 3, 1979, p. C10. Krugman's original report was presented as a paper to the annual conference (1970) of the American Association for Public Opinion Research, entitled "Electro-encephalographic Aspects of Low Involvement; Implications for the McLuhan Hypothesis."
11. On near-point vision and the incidence of dyslexia, see Barrington Nevitt's remarks on Arthur Hurst's hypothesis in *The Communication Ecology,* pp. 60–61. Near-point vision in this context refers to the propensity of some children to read with one eye only, with its possible connection to early and intense television viewing.
12. Ruth Benedict, *The Chrysanthemum and the Sword,* pp. 247–248.
13. Okakura Kakuzo, *The Book of Tea,* p. 44.
14. Kakuzo, *The Book of Tea,* pp. 44–45.
15. Kakuzo, *The Book of Tea,* p. 46.
16. Benedict, *The Chrysanthemum and the Sword,* p. 249. "On" is an obligation passively incurred (cf. p. 116.)
17. Benedict, *The Chrysanthemum and the Sword,* p. 251.
18. Benedict, *The Chrysanthemum and the Sword,* pp. 196–197.

Chapter 6. Hidden Effects

1. Chiang Yee, *The Chinese Eye,* pp. 114, 115.
2. Fritjof Capra, *The Tao of Physics* (Preface), pp. 11–12.

3. Innis, *The Bias of Communication* (regarding the Oriental circularity of time), pp. 62, 63.

4. Joseph E. Bogen, *The Human Brain*, p. 141.

5. Jacques Lusseyran, *And There Was Light*, pp. 143–144.

6. Lusseyran, *And There Was Light*, p. 32.

7. Warren Weaver, in *The Mathematical Theory of Communication* by C. E. Shannon and W. Weaver, pp. 7–8. Both Shannon and Weaver use the same depiction of a medium.

8. C. E. Shannon and W. Weaver, *The Mathematical Theory of Communication*, p. 32.

9. Mario Bunge, *Causality: The Place of the Causal Principle in Modern Science*, p. 32. Vide Aristotle, *Metaphysics*, Book I, ch. iii, 938a, b; Book V, ch. ii, and Aristotle, *Physics*, Book II, chs. iii, vii.

10. Cf. Aristotle, *Generation of Animals*, trs. A. L. Peck (Loeb Library, 1943), pp. xliv, 3. Aristotle opens Book I by presenting formal cause as "the *logos* of the thing's essence." Edition cited, London: William Heinemann Ltd.; Cambridge: Harvard University Press, 1943.

11. W. K. Wimsatt, Jr., and C. Brooks, *Literary Criticism—A Short History*, p. 709. In their footnote, the authors cite Frye's "My (Critical) Credo," *The Kenyon Review*, XII (Winter, 1951): 91–110, and add: "Archetype" borrowed from Jung, means a primordial image, a part of the collective unconscious, the psychic residue of numberless experiences of the same kind, and thus part of the inherited response pattern of the race.

12. N. Frye, *Anatomy of Criticism: Four Essays*, p. 132. Vide the discussion in *From Cliché to Archetype* by H. M. McLuhan and W. Watson (1970).

13. The book of nature: For over a thousand years, based on the book of Genesis, the West has propounded a theory of nature as one of the forms of divine revelation. There were two "books," the book of nature and the book of Scripture, parallel texts in different idioms as it were, both subject to exegesis. Shakespeare frequently alludes to this tradition of multi-level exegesis. For example, in *As You Like It* the exiled Duke remarks to his companions that words

> are counsellors
> That fleetingly persuade me what I am. . . .
> And this our life, exempt from public haunt,
> Finds tongues in trees, books in the running brooks,
> Sermons in Stones, and good in everything.
> (II.1.10–17)

The book of nature was an encyclopedia of being: only God spoke in events. In the minds of the people of the Middle Ages every event, every case, fictitious or historic, tends to crystallize, to become a parable, an example, a proof, in order to be applied as a standing instance of a general moral truth. In the same way every utterance becomes a dictum, a maxim, a text. For every question of conduct, Scripture, legends, history, literature furnish a crowd of examples or of types, together making up a sort of moral clan, to which the matter in question belongs. Cf. J. Huizinga, *The Waning of the Middle Ages,* p. 227.

14. Bunge, *Causality,* pp. 32–33.

15. Joseph Bogen in *The Human Brain,* p. 145 notes: "Although humans of any culture, so far as we know, have the potential for reading and writing, many remain non-literate and thus fall short of acquiring the most special of left-hemisphere functions. Conversely, we can readily comprehend the concept of a society in which right-hemisphere illiteracy is the rule. Indeed, our own society (admittedly complex) seems to be, in some respects, a good example: a scholastized, post-Gutenberg-industrialized, computer-happy exaggeration of the Graeco-Roman penchant for pro-positionizing."

Chapter 7. Global Robotism: The Satisfactions

1. For a précis of general economics, art, and thought of the first industrial revolution in Europe and its effects on North America (1750 to 1870), see Wallbank, Taylor, and Bailkey, *Civilization: Past and Present* (vols. I and II combined), pp. 472–500, with special reference to European population estimates, 1800 to 1900. Population size and growth projections in this book are partially based on *The Global 2000 Report to the President* (1980), produced by the Council on Environmental Quality and the Department of State, updated by the most recent world population studies (1983) done by the U.S. Bureau of Census; with occasional reference to supplementary fertility and immigration figures by the U.S. Social Security Administration (1980). Cf. *The Global 2000 Report to the President: Entering the Twenty-first Century,* 1980; vol. I (Summary), vol. II (Technical Reports), vol. III (Documentation—Global Models); *World Population, 1983, Recent Demographic Estimates for the Countries and Regions of the World,* 1983; *U.S. Population Projections for the OASDI Cost Estimates, 1980.* Population estimates for the Los Angeles and Dallas–Forth Worth areas were taken from the 1986 Consolidated Statistics for Metropolitan Areas (CSMA), U.S. Census Bureau, Washington, D.C. World population estimates were coordinated with the *United Nations World Demographic Estimates and Projections,* 1950 to 2025, as well as *Trends and Opportunities Abroad,* published by American Demographics, Inc.,

Ithaca, New York, 1988. (1986–88 demographic statements were updated with the assistance of Brad Edmondson, Senior Editor, *American Demographics* magazine.)

2. "An image is created every 1/30th of a second—the time it takes for two complete sweeps of the screen. At any one moment, however, there is never more than one dot of light glowing on the (tv) screen. We see an entire image because the brain fills in or completes 99.999 percent of the scanned pattern each fraction of a second, below our awareness. The only picture that exists is the one we complete in our brains. . . ." Television viewing is principally a right-hemisphere activity because of its audile-tactile quality. Joyce Nelson, "As the Brain Tunes Out, The TV Admen Tune In," *The Toronto Globe and Mail,* April 16, 1983.

3. Consult proceedings papers of the 1979 convention in Las Vegas (Vision '79), primarily a newspaper clipping collection entitled "Speaking of Cable," distributed by the National Cable Television Association, Washington, D.C., for future cable penetration predictions (1980–90). Marshall McLuhan attended as a principal speaker.

4. On the consumer as instant producer, cf. *The Communications Revolution and How It Will Affect All Business and All Marketing* by E. B. Weiss, esp. pp. 22–29.

5. On the impact of fiber optics as information technology, refer to *The New Television Technologies* by Lynne Schafer Gross, pp. 147–148.

Chapter 8. Global Robotism: The Dissatisfactions

1. Fossils not molecular dating of evolution says McLuhan are still the best indicators of how man created his own survival environment. Cf. Nigel Calder, *Timescale: An Atlas of the Fourth Dimension* (1983), pp. 270–271, 279–280. See the discussion on "Human Origins," pp. 241–242, and earliest agriculture, "Genes and Travelers," pp. 88–89. Also see discussion of civilization as an artifact, Simeons, A. T. W., *Man's Presumptuous Brain,* pp. 69–79, consult Lorenz, Konrad, *On Aggression,* esp. pp. 176, 210–211.

2. The West has nevertheless been more open to innovation than the East. See "Medieval Roots of Modern Technology," by Lynn White, Jr., in *Medieval Religion and Technology*, pp. 75–91, esp. pp. 78–79.

3. *Sensus communis:* the translation of all the overt senses (seeing, hearing, tasting, smelling, and touch) into a form of synesthesia. Cf. Chapter 1.

4. The United States since Monroe's time has pursued a rationalistic military policy in Central America. Note Walter La Feber's explication of the history of Bluefields on the Miskitos Coast of Nicaragua and American covert support of Juan Estrados's revolt against the incumbent ruler, Jose Santos Zelaga, with the subsequent arrival of the U.S. Marines, in *Inevitable Revolutions—The United States in Central America,* pp. 28–68. Central America is essentially tribal (right-hemisphere).

5. On African and Asian population growth, refer to broad predictions of *Global 2000,* esp. p. 1, vol. I, and p. 25 of *World Population, 1983* with special figures on international migration, pp. 45, 55–180, 185, 195–296. By the year 2020, world food production will have achieved an overall growth rate of less than 15 percent. Most of the growth will benefit countries with high per capita food consumption; cf. *Global 2000,* p. 2.

6. On Mexican population growth rates, see *World Population, 1983,* pp. 384–385, for rate of projected population estimates.

7. Cf. Christopher Lasch, *The Culture of Narcissism.* In the original version of Narcissus, Zeus makes the youth's reflection in the water appear as the face of a stranger—infatuation rather than self-love. Delusion or not, Narcissism is a "closed system," like the videogame player. Robotism tunes one to the culture.

8. Jerzy Kosinski's ironic theme in *Being There,* a novel about a man who preferred the inside world of the television set to the outside world of people. Fantasy involves only disconnected images; dreams usually are involved with working out the problems of the real world in "reel" time, like motion pictures.

9. Throughout the 1970s Marshall McLuhan classified the computer, as with any servomechanism, as a by-product of automation, which springs from the very nature of electricity. Electricity, like light, he said, illuminates everything it touches (like an electron emitting light as it moves), impelling us to become aware of the total process (circuit) at once. Cf. "Automation," *Understanding Media.*

10. The information on program monitoring vs. systems analysis came from conversations with David Curtis, Systems Engineer, *Com-Pro*/Meszaros Associates, Inc., Buffalo, New York; 1977–84.

11. Brain structure and the computer: A rough simile can be drawn between the structure of the brain, including the central nervous system, and the computer. Memory and storage could be said to perform the functions of the left hemisphere. The CPU, or central processing unit, is the right hemisphere because it "sees" the operation of the entire machine by virtue of the program. The corpus callosum, or fibrous

bridge between the left and right brain, can be compared to the regulation of input and output levels, both internally and externally.

12. The Pythagorean elite: It is but a short step from a general sense of holism (the printout as pattern) to the number mysticism of the *tetraktys*, the ancient concept of the golden section and the harmony of the spheres. If computer education remains limited, by design or neglect, to a small part of the population, systems analysts may arise as high priests of yet another wave of neo-Pythagoreans, as bellwethers of the corporate state.

13. Security and encryption: Encryption equipment, now being manufactured by IBM, Motorola, and Datotek, attached to any voice, data, or video mode, can mask or scramble transmission signals (including whole patterns of transmission) and then unmask or unscramble them at the reception point. Encryption is very expensive, roughly $5000 a line; hence, computer network transmission growth will outrun security device installation in the foreseeable future.

14. Cf. J. F. Crean, "Automation and Canadian Banking," *The Canadian Banker and the ICB Review*, vol. 85, no. 4 (July-August 1978). Cf. J. F. Crean, "The Canadian Payments System," *The Canadian Banker and ICB Review*, vol. 85, no. 5 (October 1978). On high-velocity money see James Martin, *Future Developments in Telecommuications*, pp. 243–244, 253–256.

15. J. F. Crean, "The Canadian Payments System," pp. 20–27; see p. 22 for a contrast with the U.S. payments procedure. Also, cf. J. F. Crean, "EFTS and the Canadian Payments System," *The Canadian Banker and ICB Review*, vol. 85, no. 6 (December 1978).

16. James Martin, *Future Developments in Telecommunications*, pp. 240–241, 244–246. See also J. F. Crean, "Automation and Canadian Banking," pp. 16, 20.

17. J. F. Crean, "Contrasts in the National Payment Systems," *The Canadian Banker and ICB Review*, vol. 86, no. 1 (February 1979): 19.

18. James Martin, *Future Developments in Telecommunications*, pp. 241, 243–244, 247–249, 251–252, 256–257. The inability of national states to prevent the flow of key economic, social, and military information by electronic fund transfer nets is highlighted in current computer networking research; cf. "How Do We Best Control the Flow of Electronic Information Across Sovereign Borders," *AFIPS Conference Proceedings*, vol. 48, 1979 (National Computer Conference, '79), pp. 279–282.

19. The military man in the Western world is educated to understand that he is in a constant state of warfare in which the object is to

consistently set up "boxes" of territory to defend. His psychology is essentially Euclidean, or left-hemisphere. The recent failure of most space allocation conferences underscores the current game of trying to set up direct pipelines of radio links in near space, undisturbed by the competition. Cf. *INTELSAT,* pp. 194–196, and Sandra Hochman's *Satellite Spies.*

20. In 1977, Grumman Aerospace engineers envisioned, for NASA, the placement of a vast thirty-one-ton triple-antenna space platform containing some earth station switching capacities, which could be used to set up a "national information service: for several thousand pre-designated individuals who would receive messages on two-ounce transceivers," functioning for "twenty continuous hours." See in-house publication: *Horizons,* Grumman Aerospace Corporation, Bethpage, New York (Spring 1977).

21. The area around the planet is theoretically unlimited; however, the near-term possibility of a collision between space vehicles is nevertheless very plausible. Nearly 6000 satellite objects have gone into near-earth space since 1957. By 1988, the number of man-made satellites in low orbit, irregular or otherwise, could well total over 10,000. Nuclear accidents such as COSMOS 954 (January 24, 1976), which scattered radioactive materials over hundreds of square miles of Canada's Northwest Territory, can only increase in frequency. Hence larger satellites move into higher geosynchronous orbit. See Leo Heaps's documentation in *Operation Morning Light.*

22. The problem of signal spectrum congestion is a real one today. The nations of the Northern Hemisphere, which include Japan, the USSR, Canada, Britain and the United States, are the prime users of the geostationary orbit and, as a rule, have their satellite transponders all pointed in the same direction. Cf. *Computing Canada,* Conference Issue (November 1981).

23. On continent/global satellite communications, cf. *Telecommunications: Trends and Directions,* Seminar Program, Electronic Industries Association (1981), pp. 61–68, with supplementary reference to *A History of Engineering and Science in the Bell System: (1925–1975).*

24. Global AT&T: Since 1982, the corporate logo for American Telegraph and Telephone has been a serrated globe, signaling the long-range intentions of the company. In many company ads the invention of the 256K memory chip is linked with the declaration, "The AT&T Communications network . . . delivers voice, video data, even sensory information to every corner of the world and employs every advanced technology from lightwave systems on earth to satellites in space." (Microelectronics, photonics, digital systems, software). At 315,000

people (1988), AT&T is still one of the largest and most technically complex multi-media electronic carrier corporations, excluding the seven U.S.-based "baby Bell" regional company groups.

25. The social implications of split-brain research, detailed in Chapter 4, are explored in Norman Geschwind's "Language and the Brain," *Scientific American* elaborated in (April 1972) and David Galin's "Hemispheric Specialization: Implications for Psychiatry," *Biological Foundations of Psychiatry* (1972).

26. As a good example, a public relations publication put out by New York Telephone called "Fire: The Second Avenue Story," which describes the impact on lower Manhattan of a fire in a telephone exchange (New York Telephone, 1975).

27. Special report: "Behind AT&T's Change at the Top," *Business Week* (Nov. 6, 1978), pp. 114–115. Also, effects of divestiture reorganization on economy and business, *Time,* November 21, 1983, pp. 60–74.

28. AT&T core network and divestiture: The reorganization of the 22 local operating companies into seven regional holding groups, anticipated by Marshall McLuhan in 1978, will have little immediate technical effect on the U.S. core transcontinental communications network, now dominated by AT&T. As long as the AT&T Long Distance Division, with its regional ESS-4 computer-exchange hubs (allied with critical military net systems) remains intact, the divestiture only serves to free AT&T to expand technologically into global networking and world software services. Much of the fixed assets divested domestically by the old AT&T were superannuated and returned less than 5 percent on investment annually. The development of new digital transmission techniques, more advanced than those used by the Jet Propulsion Laboratories for "imaging" on the Voyager I and II tours through the solar system (imaging: computer assisted digital return of TV and radio signals) will place AT&T in the lead in creating an intra-satellite global network outrivaling anything developed domestically by such groups as IBM, MCI and GTE. For a veiled statement of AT&T intentions see the 1983 *AT&T Annual Report* (98th Annual Stockholders Meeting), April 20, 1983, pp. 4, 21, 23.

29. Cf. Clare D. McGillem and William P. McLauchlan, *Hermes Bound: The Policy and Technology of Telecommunications,* pp. 167–168, 172–174. In addition, pp. 173, 175, 183 for details of the DATRAN failure to sustain network capability.

30. On IBM sales aggressiveness cf. *1982 Annual Report to IBM Stockholders,* April 25, 1983, pp. 10, 23, 24, 29.

31. 1982 IBM *Annual Report* (Satellite Business Systems), p. 10.

Chapter 9. Angels to Robots:
From Euclidean Space to Einsteinian Space

1. Prior explication on the functions of figure-ground, and visual and acoustic space appeared in *ETC.; A Review of General Semantics,* vol. 34, no. 2 (June 1977).

2. On Edison cf. *Encyclopedia Americana* (Int' Ed.), vol. 9 (Danby, Conn., 1979), p. 637; see also Frank L. Dryer and others, *Edison, His Life and Inventions,* Rev. ed., 2 vols. (New York, 1929).

3. On phone installation rate see *Business Week,* Industrial Edition, #2559, Nov. 6, 1978, pp. 116, 130–131.

4. Basic electronic network structure: When one makes a phone call, one is instantly (speed of light) in the middle of all the Bell Systems. In acoustic space, there is no outside or margin. A good illustration of this observable fact is documented in the activities, on an international scale, of ABC News Central (7th floor) during the recent Solidarity crisis in Poland, as reported by Robert Friedman in "Super-News: Journalism in the High Tech Mode," *Channels of Communication,* vol. 3, no. 3 (Sept.-Oct., 1983), pp. 26–30.

5. Microelectronics and leisure time: Leisure time gained in personal microcomputer networking could be used to develop data and skills for which other people would be willing to pay—paid time not "kill time."

Chapter 10. Epilogue: Canada as Counter-Environment

1. Gertrude Stein: "In the United States there is more space where nobody is than where anybody is. That is what makes America what it is." This was quoted in the official bicentennial gift to the people of the United States, *Between Friends/Entre Amis* (Toronto, 1976), pp. 4, 24.

2. *Prefaces by George Bernard Shaw,* p. 440.

3. Morton W. Bloomfield, "Canadian English and Its Relation to Eighteenth-Century American Speech," *Journal of English and Germanic Philology,* p. 59.

4. Bloomfield, "Canadian English," p. 60.

5. Stephen Leacock, *How to Write,* p. 119.

6. Lord Durham, *Report on the Affairs of British North America,* vol. II, p. 91.

7. Margaret Atwood, *Survival,* p. 60.

8. Hugh Kenner, "The Case of the Missing Face," *Our Sense of Identity,* pp. 203–208.

9. W. H. Auden, *The Dyer's Hand,* pp. 103–104.

10. All quotations from *Daisy Miller* are taken from *The Novels and Tales of Henry James* (vol. 18), pp. 3–93, *passim.*

11. Marius Bewley, *The Complex Fate,* p. 57.

12. Hamlin Garland, *Roadside Meetings,* p. 461.

13. Donald Creighton, *The Empire of the St. Lawrence,* p. 1.

14. Ramsay Cook, *The Maple Leaf Forever,* pp. 111–112.

15. Kenneth McNaught, "Canadian Independence, Too, Was Won in the 1770's," *Toronto Star* (1976), p. C3.

16. T. S. Eliot, "American Literature and the American Language," *To Criticize the Critic,* p. 45.

17. Eliot, "American Literature," pp. 44, 54.

18. Frederick J. Turner, *The Frontier in American History,* pp. 2–3.

Selected Bibliography

AFIPS Conference Proceedings (National Computer Conference 1979).
Vol. 48, May 1979.

Aristotle. *De Anima*. *Translated* by W. D. Ross. Oxford: Clarendon
Press, 1931.

———. *Generation of Animals*. Translated by A. L. Peck. Cambridge, Mass.: Harvard University Press, 1943.

———. *The Rhetoric of Aristotle*. Translated by Lane Cooper. New
York: Appleton-Century-Crofts, 1960.

AT&T Annual Report, 1983 (Annual Report to Stockholders). Charles
L. Brown, Chairman of the Board, April 20, 1983.

Atwood, Margaret. *Survival*. Toronto: House of Anansi Press, 1972.

Auden, W. H. *The Dyer's Hand*. New York: Random House, 1962.

Benedict, Ruth. *The Chrysanthemum and the Sword, Patterns of Japanese Culture*. Cleveland and New York: Meridian Books, 1967.

Between Friends/Entre Amis. Toronto: McClelland and Stewart, 1976.

Bewley, Marius. *The Complex Fate*. London: Chatto and Windus,
1952.

Birdsall, Derek, and Carlo M. Cipolla. *The Technology of Man, A
Visual History*. London: Wildwood House, 1980.

Bloomfield, Morton W. "Canadian English and Its Relation to Eighteenth-Century American Speech." *Journal of English and Germanic Philology* 47 (1948).

Bogen, Joseph E. (M.D.) and Glenda M. Bogen (R.N.). "The Other
Side of the Brain, III: The Corpus Callosum and Creativity."
Bulletin of the Los Angeles Neurological Societies 34, no. 4
(October 1969).

Bunge, Mario. *Causality: The Place of the Causal Principle in Modern
Science*. Cambridge: Harvard University Press, 1959.

Calder, Nigel. *Timescale: An Atlas of the Fourth Dimension.* New York: Viking Press, 1983.

Capra, Fritjof. *The Tao of Physics.* Boulder, Colo.: Shanghala Publications, 1975.

Carpenter, Edmund Snow. *Oh, What a Blow That Phantom Gave Me!* New York: Holt, Rinehart and Winston, 1973.

Cicero, Marcus Tullius. *De Oratore.* Translated by E. W. Sutton and H. Rackham. Cambridge, Mass.: Harvard University Press, 1967.

Cook, Ramsay. *The Maple Leaf Forever.* Toronto: Macmillan, 1971.

Cooper, Leon, and Michael Imbert. "Seat of Memory." *The Sciences* (February 1981).

Cooper, S. F., Jr. *Imaging Saturn: The Voyager Flights to Saturn.* New York: Holt, Rinehart and Winston, 1983.

Cornford, F. M. "The Invention of Space." *Essays in Honour of Gilbert Murray.* London: Allen and Unwin, 1936.

Crean, J. F. "Automation and Canadian Banking." *The Canadian Banker and the ICB Review* 85, no. 4 (July-August 1978).

———. "The Canadian Payments System." *The Canadian Banker and the ICB Review* 85, no. 5 (October 1978).

———. "Contrasts in National Payment Systems." *The Canadian Banker and the ICB Review* 86, no. 1 (February 1979).

———. "EFTS and the Canadian Payments System." *The Canadian Banker and the ICB Review* 85, no. 6 (December 1978).

Creighton, Donald. *The Empire of the St. Lawrence.* Toronto: Macmillan, 1956.

Lord Durham. *Report on the Affairs of British North America.* Oxford: Clarendon Press, 1912.

Edwards, Betty. *Drawing on the Right Side of the Brain.* Los Angeles and New York: J. P. Tarchner; St. Martin's Press, 1979.

Eliot, T. S. "American Literature and the American Language," in *To Criticize the Critic.* New York: Farrar, Straus and Giroux, 1965.

———. *Selected Essays.* London: Faber and Faber, 1932, 1950.

Ellul, Jacques. *Propaganda: The Formation of Men's Attitudes.* Translated by Konrad Keller and Jean Lerner. New York: Vintage Books, 1973.

Eysenck, H. J. "Personality and the Law of Effect," in *Pleasure, Reward Preference.* Edited by D. E. Berlyne and K. B. Madsen. New York and London: Academic Press, 1973.

Fagan, M. D., ed. *A History of Engineering and Science in the Bell*

System (1925–1975). Murray Hill, New York: Bell Telephone Laboratories, 1978.

Fincher, Jack. *Human Intelligence.* New York: Putnam, 1976.

Friedman, Robert. "SuperNews: Journalism in the High Tech Mode." *Channels of Communication* 3, no. 3 (Sept.-Oct., 1983).

Frye, Northrop. *Anatomy of Criticism: Four Essays.* Princeton, N.J.: Princeton University Press, 1957.

Galin, David. "Hemispheric Specialization: The Implications for Psychiatry," in *Biological Foundations of Psychiatry.* Edited by R. G. Grenell and S. Gabay. New York: Raven Press, 1976.

Garland, Hamlin. *Roadside Meetings.* New York: Macmillan, 1930.

"Geostationary Congestion." *Computing Canada* (Conference Issue). November 1981.

Geschwind, Norman. "Language and the Brain," in *Scientific American* 226 (April 1972).

Giedion, Siegfried. "Space Conception in Prehistoric Art," in *Explorations in Communications, An Anthology.* Edited by Edmund Carpenter and Marshall McLuhan. Beacon Press, 1966.

———. *Space, Time and Architecture,* Cambridge, Mass.: Harvard University Press, 1941.

The Global 2000 Report to the President: Entering the Twenty-first Century. Vols. I, II, and III. Gerald O. Barney, Study Director. Washington, D.C.: U.S. Government Printing Office, 1980.

Gombrich, E. H. *Art and Illusion.* Princeton: Princeton University Press, 1961.

Gross, Lynne Schafer. *The New Television Technologies.* Dubuque, Iowa: Wm. C. Brown, 1983.

Gusdorf, Georges. *Speaking (La Parole).* Translated by Paul T. Brockelman. Evanston, Ill.: Northwestern University Press, 1965.

Hass, Hans. *The Human Animal.* New York: G. P. Putnam, 1970.

Havelock, Eric. "Origins of Western Literacy," in *Ontario Institute for Studies in Education.* Monograph series no. 14. Toronto: 1971.

———. *Preface to Plato.* Cambridge, Mass.: Harvard University Press, 1963.

Heaps, Leo. *Operation Morning Light (COSMOS 954).* New York: Paddington Press, 1978.

Heraclitus. *The Cosmic Fragments.* Translated by G. S. Kirk. Cambridge: Cambridge University Press, 1962.

Hochman, Sandra, and Sybil Wong. *Satellite Spies.* New York: Bobbs-Merrill, 1970.

Huizinga, J. *The Waning of the Middle Ages.* New York: Doubleday Anchor Books, 1955.

1982 IBM Annual Report (Annual Report to Stockholders). Frank T. Carey, Chairman of the Board, April 25, 1983.

Innis, Harold. *The Bias of Communication.* Toronto: University of Toronto Press, 1951.

―――. *Empire and Communications.* London: Oxford University Press, 1950.

James, Henry. *The Novels and Tales of Henry James.* New York: Charles Scribner's Sons, 1909. Reprint. New York: Augustus M. Kelley, 1972.

Jung, Carl G. *Psyche and Symbol.* New York: Doubleday Anchor Books, 1958.

Kakuzo, Okakura. *The Book of Tea.* Rutland, Vermont–Tokyo, Japan: Charles E. Tuttle, Co., 1978.

Kenner, Hugh. "The Case of the Missing Face," in *Our Sense of Identity.* Edited by Malcolm Ross. Toronto: Ryerson Press, 1954.

Kirk, Geoffrey S. *Heraclitus: The Cosmic Fragments.* London: Cambridge University Press, 1954.

Kohler, Wolfgang. *Gestalt Psychology.* New York: Liveright, 1947.

Kosinski, Jerzy. *Being There.* New York: Harcourt Brace Jovanovich, 1971.

Krugman, Herbert E. "Electro-encephalographic Aspects of Low Involvement, Implications for the McLuhan Hypothesis," presented at the meeting of the American Association of Public Opinion Research, Lake George, New York, May 21–23, 1970.

Kuhn, Thomas. *The Structure of Scientific Revolutions.* Chicago: University of Chicago Press, 1962.

La Feber, Walter. *Inevitable Revolutions: The United States in Central America,* New York: W. W. Norton, 1983.

Lain, Entralgo Pedro. *Therapy of the Word in Classical Antiquity.* New Haven: Yale University Press, 1970.

Lasch, Christopher. *The Culture of Narcissism.* New York: W. W. Norton, 1973.

Leach, Edmund Ronald. *Claude Lévi-Strauss.* New York: Viking Press, 1970.

Leacock, Stephen. *How to Write.* New York: Dodd, Mead, 1943.

Lewis, Wyndham. *The Caliph's Design: Architects Where Is Your Vortex?* London: Egerest Press, 1919.

Lilly, John C. *The Centre of the Cyclone.* New York: Bantam Books, 1973.

Lin, Nan. *The Study of Human Communication.* Indianapolis: Bobbs-Merrill, 1973.

Lorenz, Konrad. *On Aggression.* Translated by Marjorie Kee Wilson. New York: Harcourt, Brace, World, 1966.

Luria, A. F. "The Functional Organization of the Brain," in *Scientific American* (March 1970).

————. *The Working Brain.* Translated by Basil Haigh. New York: Basic Books, 1973.

Lusseyran, Jacques. *And There Was Light.* Translated by Elizabeth Cameron. Boston: Little, Brown, 1963.

McGillem, Clare D., and William P. McLauchlan. *Hermes Bound: The Policy and Technology of Telecommunications, Science and Society.* Purdue University Series in Science, Technology and Human Values. West Lafayette, Ind.: Purdue University Press, 1978.

McLuhan, Marshall, with Wilfred Watson. *From Cliché to Archetype.* New York: Viking Press, 1970.

————. *The Gutenberg Galaxy.* Toronto: University of Toronto Press, 1962.

McNaught, Kenneth. "Canadian Independence, Too, Was Won in the 1770's," *The Toronto Star,* July 1, 1976, p. C3.

Marrou, Henri Irenee. *A History of Education in Antiquity.* Translated by Charles Lamb. New York: Sheed and Ward, 1956.

————. *Saint Augustin et la fin de la culture antique.* Editeur: E. De Boccard. Paris: Bibliotheque des Ecoles Francaises D'Athenes et Rome, 1938.

Martin, James. *Future Developments in Telecommunications.* Englewood Cliffs, N.J.: Prentice Hall, 1977.

Nelson, Joyce. "As the Brain Tunes Out, the TV Admen Tune In." *The Globe and Mail,* Toronto, April 16, 1983.

Nevitt, Barrington. *The Communication Ecology.* Toronto: Butterworth and Company, 1982.

Nin, Anais. *D. H. Lawrence.* Chicago: Swallow Press, 1964.

Ong, Walter. *Ramus, Method and Decay of Dialogue.* Cambridge, Mass.: Harvard University Press, 1958.

Ornstein, Robert. *The Psychology of Consciousness.* New York: Penguin Books, 1972.

Pauling, Linus Carl. *The Nature of the Chemical Bond and the Structure of Molecules.* Ithaca, N.Y.: Cornell University Press, 1939.

Penfield, Wilder. "Memory Mechanisms," *A.M.A. Archives of Neurology and Psychiatry* 67 (1952).

Piaget, Jean. *Structuralism*. Translated and edited by Chaninah Maschler. Harper Torchbooks, 1971.

Popper, Karl. *The Open Society and Its Enemies*. Princeton, N.J.: Princeton University Press, 1966.

Powell, James N. *The Tao of Symbols*. New York: Quill/William Morrow, 1982.

Reich, Charles A. *The Greening of America*. New York: Random House, 1970.

Restak, Richard M. (M.D.). *The Brain: The Last Frontier*. New York: Warner Books, 1979.

Ricoeur, Paul. *The Rule of Metaphor*. Toronto: University of Toronto Press, 1977.

Rosenstock-Huessy, Eugen. *Out of Revolution: Autobiography of Western Man*. Norwich, Vt.: Argo Books, 1969.

Ross, Frank, Jr. *Oracle Bones, Stars and Wheelbarrows: Ancient Chinese Science and Technology*. Boston: Houghton, Mifflin, 1982.

Saussure, Ferdinand de. *Course in General Linguistics*. Edited by Charles Bally and Albert Sechehaye in collaboration with Albert Riedlinger. Translated with introduction and notes by Wade Baskin. New York: McGraw-Hill, 1959.

Schwartz, Tony. *The Responsive Chord*. Garden City, N.Y.: Doubleday Anchor Books, 1973.

Semmes, Josephine. "Hemispheric Specialization: A Clue to Possible Mechanism." *Neuropsychologica* 6 (1968).

Shannon, C. E., and W. Weaver. *The Mathematical Theory of Communication*. Urbana, Ill.: University of Illinois Press, 1963.

Shaw, George B. *Prefaces by George Bernard Shaw*. London: Constable and Company, 1934.

Simeons, A. T. W. *Man's Presumptuous Brain: An Evolutionary Interpretation of Psychosomatic Disease*. New York: E. P. Dutton, 1962.

Slater, Philip E. *The Glory of Hera: Greek Mythology and the Greek Family*. Boston: Beacon Press, 1968.

Striegel, James F. "Marshall McLuhan on Media." Rockville, Md. (Unpublished dissertation for the Goodwin Watson Institute/Union Graduate School, 1978).

Telecommunications: Trends and Directions. A seminar program sponsored by the Communication Division, Electronic Industries Association (proceedings editor: John Sodolski), May 26–28, 1981, Hyannis, Mass.

Trotter, R. J. "The Other Hemisphere." *Science News* 109 (April 3, 1976).

Tucker, Albert W., and Herbert S. Bailey, Jr. "Topology." *Scientific American* 182, no. 1 (January 1950).

Turbayne, C. M. *The Myth of Metaphor*. New Haven: Yale University Press, 1962.

Turner, Frederick Jackson. *The Frontier in American History*. New York: Henry Holt, 1920.

U.S. Population Projections for the OASDI, Cost Estimates, 1980. Social Security Administration Publications, no. 47061.82. Francisco R. Bayo and Joseph F. Faber. Washington, D.C.: U.S. Government Printing Office (H.E. 3.19:82), June 1980.

Wallbank, Walter T., Alistair M. Taylor, and Nels M. Bailkey. *Civilization: Past and Present* (Single Volume Edition). Chicago: Scott, Foresman, 1962.

Weiss, E. B. *The Communications Revolution and How It Will Affect All Business and All Marketing* (Advertising Age Monograph). Chicago: Advertising Publications, 1967.

White, Lynn, Jr. *Medieval Religion and Technology, Collected Essays*. Center for Medieval and Renaissance Studies. Los Angeles: University of California, 1978.

Wimsatt, W. K., Jr., and Cleanth Brooks. *Literary Criticism—A Short History*. New York: Alfred A. Knopf, 1957.

Wittrock, M. C., ed. *The Human Brain*. Englewood Cliffs, N.J.: Prentice Hall, 1977.

World Population, 1983: Recent Demographic Estimates for the Countries and Regions of the World. Bureau of the Census. C. E. Kincannon, Dir. U.S. Department of Commerce. Washington, D.C.: U.S. Government Printing Office, Dec. 1983.

Yee, Chiang. *The Chinese Eye: An Interpretation of Chinese Painting*. Bloomington: Indiana University Press, 1964.

Acknowledgments

The Global Village could not have been finished without the active support of Mrs. Corinne McLuhan and Mrs. Matie Molinaro (agent and literary executor of the McLuhan Estate), who together provided access to the McLuhan Papers and approval of the text. Additional assistance was also forthcoming from Joseph Keogh, a former research assistant of Marshall McLuhan and George Thompson, Marshall's longtime administrative aide at the Centre for Culture and Technology until 1980. Marsha Seifert of the University of Pennsylvania gave much appreciated editorial assistance.

Technical details were provided to me on occasion by Robert Hinkleman of AT&T Long Lines Division. In earlier stages of the manuscript, Gordon Thompson, senior scientist at Bell-Northern Research, Ottawa, gave Marshall McLuhan much technical advice. The late artist, York Wilson, generously commented on Marshall's past collaboration with Harley Parker and Wilfred Watson. Graphics were done by Blair Schrecongost. Barrington Nevitt meticulously read Chapters 1 and 3 and allowed me to utilize much of the research data he had developed with the McLuhans. Of special help was David Curtis, computer systems engineer of *Com-pro* Consulting Service, Buffalo, New York. My thanks also to Professor Neil Postman and Dr. George Gerbner who published excerpts from the preliminary draft in *ET CETERA*, a review of General Semantics, and the *Journal of Communication*. Finally, my thanks to Scott Lenz of Oxford University Press who copyedited the working manuscript with great good humor and scholarly finesse.

Index

left/right brain (*Cont.*)
  working, 119–20, 126, 127–28,
  140–41, 142; and grammar, 33,
  34, 64; and Greek culture, 33,
  61–62; and hardware/software,
  128; and hearing, 52; and hierarchy, 35, 48, 52; and history,
  40; and holism, 59–60; and
  identity, 73–74, 99–100; and individualism, 65, 126; and information processing, 53–54; interaction between the, 62; inter-
  changeability of, 8; and labeling,
  50, 54; and language, 7, 23, 30–
  31, 38, 50, 52, 53–54, 55; and
  learning disabilities, 64; and the
  left/right side of the body, 52;
  and linearity, 52, 54, 58, 63–64,
  74, 121; and literacy, 61–62, 64;
  and logic, 121; and *logos,* 63–64;
  and mathematics, 58; and metaphors, 28, 29–30, 52; multi-
  dimensional aspects of, 50–51;
  and myth, 35; and objectivity,
  74–75; and oral cultures, 56; and
  the Oriental culture, 56, 62–63,
  65–66, 68, 71, 72–73; and pattern recognition, 38; Plato's views
  about, 57–58; and print, 63–64;
  and the quality of life, 101; and
  quantification, 21–22, 50; and
  rationalism, 61–62, 121; and
  relationships, 65–66; and the
  resonant interval, 63–64; and
  rhetoric, 33, 34; and science, 21–
  22, 58, 72–73; and sensibilities,
  74–75; and sequentiality, 50, 52,
  54, 58, 73, 74; and the shift in culture, ix, x, 35, 45, 80, 83–87; and
  simultaneity, 40, 54–55, 56, 59,
  62, 63–64, 65; spatial property
  of, 48, 52, 53–54, 59; and specialism, 59–60, 67, 74; and
  speech, 23, 52, 55, 73; and stroke
  victims, 73; suppression of the,
  62; and tactile space, 35, 48, 52,
  53–54, 59; and technologies, 3–4,

63–64, 73–74, 147; and the
  tetrad, 3–4, 6, 7, 102–3; and
  time, 10, 73–74; and the triad,
  6–7; and tribalism, 65; and
  Western culture, 13, 21, 48, 49,
  54, 62, 63–64, 66, 71–73, 74,
  100–101, 133; and words, 7; and
  writing, 52, 73
left side of the body, 52
leisure time, 114–15, 143
Lévi-Strauss, Claude, 23–24, 25
Lewis, Wyndham, 66
liberal education, 33–34
linearity: and acoustic/visual space,
  36–39, 45, 59; and the alphabet,
  58, 73; and angelism, 69–70; and
  causality, 77; and communication
  theories, 80; and the electronic
  age/technologies, 58, 68, 73, 75;
  and Euclidean geometry/space,
  55, 135; and figure-ground relationship, 75; and language, 49;
  and rationalism/science, 58, 73;
  and stroke victims, 73; and the
  left/right brain, 52, 54, 58, 63–
  64, 74, 121; and Western culture,
  54, 58, 68, 73, 75, 77, 80, 86,
  133–34. *See also* time
linguistics, x–xi, x, 22, 23, 27, 30,
  45–46
literacy, 60, 61–62, 64, 68, 93, 99,
  135, 137
literary criticism, 30, 78–79
logic, 6–7, 39, 80, 107, 121
logos, 7, 30–33, 34, 36, 63–64, 78
long lines, 125, 128, 141
loyalties, 98
Lucretius, 132
Luria, A. R., 58, 73
Lusseyran, Jacques, 27–28, 35–36,
  37, 74–75

MacLean, Paul, 52
McNaught, Kenneth, 161
Malinowski, Bronislaw, 40
ma (negative space), 12, 39, 85
man as god, 3, 97–98